TECHNOLOGY

EPISTEME

A SERIES IN THE FOUNDATIONAL,
METHODOLOGICAL, PHILOSOPHICAL, PSYCHOLOGICAL,
SOCIOLOGICAL, AND POLITICAL ASPECTS
OF THE SCIENCES, PURE AND APPLIED

Editor: MARIO BUNGE

Foundations and Philosophy of Science Unit, McGill University

Advisory Editorial Board:

VOLUME 11

JOSEPH AGASSI

Tel-Aviv University and York University, Toronto

TECHNOLOGY

Philosophical and Social Aspects

D. REIDEL PUBLISHING COMPANY

A MEMBER OF THE KLUWER ACADEMIC PUBLISHERS GROUP

DORDRECHT / BOSTON / LANCASTER / TOKYO

Library of Congress Cataloging in Publication Data

Agassi, Joseph.
Technology, philosophical and social aspects.

(Episteme; v. 11)
Includes index.
1. Technology--Philosophy. 2. Technology--Social aspects.
I. Title. II. Series: Episteme (D. Reidel); v. 11.
T14.A33   1985   303.4'83   85-14228
ISBN 90-277-2044-4
ISBN 90-277-2045-2 (pbk.: Pallas edition)

Published by D. Reidel Publishing Company,
P.O. Box 17, 3300 AA Dordrecht, Holland.

Sold and distributed in the U.S.A. and Canada
by Kluwer Academic Publishers,
190 Old Derby Street, Hingham, MA 02043, U.S.A.

In all other countries, sold and distributed
by Kluwer Academic Publishers Group,
P.O. Box 322, 3300 AH Dordrecht, Holland.

For Pozzi Escot and Robert Cogan

TABLE OF CONTENTS

PART TWO: TECHNOLOGY AND PHILOSOPHY

EDITORIAL PREFACE

This is a distinguished contribution to the philosophy and sociology of technology by one of the pioneers in these fields. The latter are comparatively new and they are not being cultivated vigorously enough. This is surprising given that technology, together with capital, was the motor of the industrial revolution that started in mid 18th century and that may never be completed.

The neglect of the philosophy and sociology of technology seems to be due to three major factors. One is that many scholars confuse technology with science, so that when dealing with either of them they believe to have taken care of the other as well. A second reason is that most scholars do not realize the conceptual richness of technology: they do not understand that, unlike the traditional crafts, modern technology presupposes science and involves research, design, and planning, all of which pose intriguing conceptual problems. A third reason for the neglect is the traditional contempt of the scholar for everything that smells of manual work.

This situation has started to change over the past few years, partly under the influence of Professor Agassi's numerous writings and those of a few other scholars who have explained that technology presupposes and raises a number of interesting philosophical problems, and that technologists, unlike basic scientists, are accountable to both their employers and the public at large. In fact, the philosophy and sociology of technology are expanding quickly. There are professional societies and periodicals devoted to them, and an estimated 2000 courses on science, technology and society are currently being taught around the world.

A major problem faced by any teacher or student of a course in the philosophy and sociology of technology is the dearth of good textbooks on the subject. The present work, a product of two decades of research and teaching on three continents, is a suitable textbook for any course on the philosophical and sociological aspects of technology. It covers an extensive ground in a clear and concise manner, and without using professional jaegon.

Agassi's book gives us a faithful and clear picture of contemporary technology as both a product of human ingenuity and a powerful means for altering the world – for better or worse. It is also an eloquent plea for the

xi

democratic control of technology, a cultural force that, though ambivalent, is never socially neutral.

Foundations & Philosophy of Science Unit,   MARIO BUNGE
McGill University, Montreal

PREFACE

The progress of man by the
education of the mind – there
is no safety but in that.
Victor Hugo
"The Mind and the Masses"

All societies have technology and control it by diverse means and with the
aid of diverse social and political institutions. The eighteenth century thinkers of the Enlightenment movement considered technology as a peculiarly
high form of applied natural science and disregarded all tradition, including,
incidentally, many traditional social controls of technology. Adam Smith
expressed the spirit of the age when he said, enterpreneurs should control
the machines they own, not the government, the control of production as a
whole should be effected by the individual consumers through the open
market. In the nineteenth-century, thinkers of the Reaction to the Enlightenment movement emphasized the ill effects which industrialization causes,
and called for the maintenance of traditions and of communal life. Karl
Marx expressed the view, which soon became more influential than those
of most modern thinkers, that the technological stage of development of a
society is the sole basic determinant of its social and political structures. In
recent years a new view is emerging, to the elaboration of which the present
volume is devoted, which present social and physical technology as strongly
interacting to varying degrees of satisfactoriness. To achieve a satisfactory
man-machine integration we need a new technology – which should coordinate and harmonize social and physical technology.

Technologies, especially agrarian, have destroyed societies which could
not control them well enough. Today technology threatens to destroy the
human race. This is why the task of the new technology is both so important
and so urgent.

The bias of the present book is frankly political: we have to implement
a drastic change in our policy towards the implementation of technology,
and center less on the physical and more on the social side of technology:
we should prefer the change of the organization of a system to the intro-

xiii

duction of a new piece of machinery. And we should study the social side-effects of technological innovation and decide whether they are desirable or not, and if not, what to do about it. Such decisions are political, and the chief political question is, what political machinery should exersize such controls.

The political bias of the present book is frankly democratic: its opts for democratic control and, moreover, for a broad public participation in the political process. The major task thus posed to us, before it is too late – if it is not too late already – is that of democratization.

The democratization bias of the present book is, finally, educational: there is no short-cut that circumvents education. Indeed, the major contribution of technology from time immemorial is that of creating more leisure time and more educational opportunities.

All anti-technological schools of thought are here opposed as reactionary and as impractical. We need new technologies to help us fight the ills of the extant technologies. We can use the means of mass communication to create mass movements and use these as means for rapid mass education for democracy and as means for pressure on legislatures to enact new means of democratic controls of technologies.

Once we see democracy as the process of increased participation of citizens in the political process through education, we cannot fail to realize that the dissemination of political knowledge and information is essential to democracy and so calls for institutionalization – of both the dissemination of information and of democratic control of it. This leads us to perceive at once the terrible crisis in contemporary political life, the credibility-gap so-called, the popularity of the view that one cannot succeed in politics without being a liar. The credibility-gap is a top priority target for all those who cherish democracy and who deem the future of humanity dependent on the survival of the democratic way of life.

So much for the bias of this book. The discussion in it pertains to this bias and to its background, historical and philosophical, and airs questions whose inadequate study in our society gives democracy less than the credit it merits. The major question, however, pertains to the organization of institutions which handle technology on a large scale.

To say what the field of the organization of technology is, may be to utter a prejudice concerning questions within this field. One might, to avoid complications from the start, give a coarse description of the field and the major activities current in it, and then move towards one's own more refined view, and the prejudices it includes may then be a bit more transparent than if the crude description is avoided.

Crudely, current philosophy of technology regrettably debates not technological matters but technological society. It hardly notices questions such
as, When is technology scientific? or, To what extent does technology
depend on natural science? or, What is the social organization best
conducive to the solution of certain major technological problems? Rather,
it centers on the question, Is technology good or bad for humanity? Have
we lost control over our machines? What is the political ideology which
makes technology bloom?

Current philosophy of technology largely divides into the pro and the anti,
the optimists and the pessimists. For my part, I regard both schools as
pernicious, since they engage in a scholastic question. Clearly, mankind is
not able or willing to give up absolutely all techniques, and clearly some
techniques are forbidden in every society. The question is, then, where and
how to draw the line?

There are, by and large, two (allegedly) progressive dominant schools
about these matters, both (unintentionally) highly reactionary, both preaching the politics of reaction based on the philosophy of political impotence
which rests on the speculative idea of social or technological historical
determinism or the doctrine of historical inevitability. The one school
preaches the (allegedly) capitalist system as the chief cause for the beneficial
technological boom. It favors the maximal support for free, competitive,
market economy. People holding this view are known as right-wing radicals.
The others preach the opposite, (allegedly) Marxist ideology. They are
known as left-wing radicals. Both parties declare unavoidable the concentration of immense political influence through the concentration of wealth,
in what is known as the military-industrial complex and the multi-national
corporations – although one party adores and the other loathes the big
corporations which allegedly run the modern industrialized world, and
which are obviously not built for political action and which interfere in
political affairs – often detrimentally, out of short-sighted considerations.

The two schools of thought are very important in that they are nourished
by powerful traditional ideas of the modern world, they are very popular, and
they constitute a positive danger to the world at large, and may very well be
instrumental in the process of the destruction of humanity in the near future.
For, no doubt, the anti-technological school, the pessimists, are quite right
when they remind us of dangers inherent in the division of the world to rich
and poor nations, the pollution of the environment, the populations explosion, especially in the poor countries, and the proliferation of nuclear
weapons, especially to dictatorships of oil-rich poor countries (the four P's:
Poverty, Pollution, Population-explosion, and Proliferation of weapons).

Unless something is done about these dangers, and on a global scale, effectively, and soon and fast, we may very well be doomed. And the current political ideological debate on the matters at hand pollutes the intellectual space – perhaps enough to ensure failure.

So much for the coarse picture. When we go to further detail we find that economists are almost all on the side of the radical right. This might be masked by the fact that there is in the west a powerful school of economists who oppose the radical right – the Keynsians – who are, it seems, a-little-less-radical-than-the-radical-right. But they all endorse the ideal of the open market, the theory of consumers' preference of one sort or another, and of similar allegedly individualistic tendencies. In this book I will therefore attack these economic ideas, not the left-wing ones, which are hardly articulated vague notions anyhow, and when one tries to articulate them they sound so fantastically romantic one cannot trust one's articulations. There is no left-wing economic theory proper, anyway.

The problem of world safety is organizational, and, in organizational theory jargon, it is the problem of balance between maintenance and adjustment. For the sake of its maintenance mountains can be moved. It is not standing still, but it moves with minimal adjustments, so that when it moves off course much is required to effect even a small adjustment. The system does adjust, but with great effort and while greatly risking its maintenance. The same holds for the big corporations, and more so for all military organizations. The military industrial complex is thus also on balance more maintenance than adjustment. Yet corporations compete and so are forced to adjust more than a socialist government. This is the point illustrated by the success of Japanese industry, which is much less individualistic than the western one, yet has achieved more adjustment than balance. Japan has developed the most un-Japanese institutionalized techniques of publicly insulting its unsuccessful administrators, among other means.

This is not the place to discuss the possibility of a better balance between maintenance and adjustment; suffice it to observe that an industrial organization is better adjustable in a mixed economy than in either extreme system, and that the maintenance desperately needed today is not of any specific organization but of the ecosystem – spaceship earth has to be maintained, as Buckminster Fuller said. This is the main topic of every philosophy of technology worth its name.

In this study, then, most of the attention will be given to an overview and to the general outline of how a middle-view may be forged that avoids both the optimist-pessimist controversy over technology and all radical views of economics – right and left. And the chief objection offered here is that they are all inherently anti-political and preaching political inaction; they are thus expressions of helplessness and of anti-democratic tendencies. Therefore, the argument in this study is to a large extent broad and general – avoiding technicalities whenever possible, and exploring those which could not be avoided, whether from the field of the theory of science – metascience – or of political theory. At times economic arguments too could not be avoided, but they are chiefly meant to expose the pseudo-scientific parts of radical-right, neo-classic economic theory, which are the staple diet of all elementary economics training, at least in the industrialized world. This book does not pretend to contribute to economic theory but to support the "big-push" theory of economic development, so-called, and place it in a reasonably comprehensive intellectual framework. In economics the radical-right, neo-classic theory is a sinking ship; but we have no time to wait till the last rats have left it, since they have incentives to stay on board, employed as they are as economic advisers and public relations officers and media-apologists for the thoughtlessness that characterizes international politics these days.

The time is short and the task is vast.

ACKNOWLEDGEMENTS

Judith Buber Agassi had many discussions with me on the topics of this book and on related topics. She taught me sociology of work and I ventured to make an integral part of the outlook presented here. She has also, as have Mario Bunge and I. C. Jarvie, read the penultimate draft and made many encouraging comments, corrections and suggestions. Mark A. Greenbert went over the draft and made very valuable editorial corrections and suggestions.

My profound gratitude to them all.

Chapter 3, Section 3, is based on the article "Cybernetics" in the *Hebrew Encyclopedia,* which was published unsigned at my insistence: an editor of that encyclopedia had inserted in that essay, without bothering to procure the author's consent, an uninformative quotation from a famous philosopher. (Regrettably references to famous thinkers are often used as mere means of legitimation.)

Chapter 9 was previously published in *Methodology and Science* **17,** 1984, pp. 1–24.

Chapter 10 was previously published in *Research in Philosophy and Technology* **7,** 1984, pp. 193–210.

Chapter 12, Section 3 is based on a paper in the same annual, **1,** 1978, pp. 53–64.

My gratitude to the editors of these periodicals for permission to reprint.

INTRODUCTION

This book is written in order to deliver to the reader an urgent message, rooted in the following well-known facts. (1) Modern technology has created the means by which mankind can destroy life on earth, or at least bring an end to human existence as known today. (2) Most people, including mankind's current political leadership, ignore this fact; most of the rest are trying to force this fact into the political agenda by one and only one means: the creating and spreading of a sense of panic. Now this means will not do: the panic leads to hostility to technology, yet we need more technology, not less. What we need urgently, then, is the invention and development and implementation of means of control – which is to say a technology – so as to control technology to the extent required for the prevention of the catastrophes it may bring about – pollution, population explosion, and nuclear devastation. And, no doubt, by its nature, such technology, such means of control of technology at large, is political. The bias of the present book is thus political. And within politics, the bias in this book is democratic.

In brief, this book is meant to be one small addition to the campaign for the intensification of the search for better tools for the democratic control of technology.

1. CONTROL IN GENERAL

People do things with or without awareness and forethought, with or without training, along specific or general lines of actions. The general line of action that we can discern, we call a technique; the theory of techniques is called technology. The philosophy of technology is a collection of attitudes towards technology, the evaluations of technology – in general or in particular – and the search for criteria for such evaluations, as well as of problems emerging from the presence of technology in our midst, and most particularly the classical Golem problem, which is, how can we avoid losing control over our machines? This problem is an example of a very simple technological problem generalized and thus made philosophical. An engineer may very well worry about controls over this or that piece of machinery. A statesman may very well worry about controls over his generals. Goethe, that celebrated

1

poet, worried about control over his vocabulary, which worry he symbolized by the fable of the sorcerer's apprentice, popularized by Paul Ducas and Walt Disney in *Fantasia*. Unlike the general problem of control which is a problem of philosophy, each of the other problems mentioned presents a specific technical problem that can be solved with specific techniques or methods or ways. Controlling a machine and controlling a war-machine – a general and his men – may be vastly different, and so it is useful to place one under the heading of physical technology and the other under the heading of political technology. We may also use the labels of biotechnology and of social technology and more, including genetic technology and the technology related to artistic techniques – including techniques by which a writer should secure his control over his vocabulary, which in itself is really fairly easy except that authors often lose their wish to control their vocabulary and enjoy the state of folly where words take over and cease to be mere vehicles for deliberation or mere artistic building blocks. This, however, also calls for some psychological technology. There are, of course, known psychological techniques – whether techniques for self-control, for maintaining peace of mind, or for the prolongation of moments of delight of all sorts. Whether these techniques will develop into technologies is a complex matter of specific conditions, of specific accumulations of techniques, or of specific philosophies, perhaps.

An important disclaimer might be offered first. When we view techniques as means of control of the environment, usually in the hand of the technician or engineer, and when we view technology as the study and control of the techniques which are in the hand of the engineer, then one might at once conclude that since theory should guide practice, the engineer should by right be the boss of the technician. By analogy, then, the philosopher of technology is the natural candidate for the role of one who aspires to be the boss of engineers. That philosophers do aspire to be bosses is generally admitted – and this means not only Plato's legendary philosopher king but also many philosophers throughout history and also many more of our colleagues and acquaintances who actively seek power. But some of us think bossing always wrong. Thus, it is not in the least the right of engineers to boss technicians; even in factories and plants where engineers belong to management and do boss technicians, the situation may be judged unhealthy, and for good reasons. Within philosophy, the view of a human hierarchy parallel to levels of theorizing simply cannot obtain: theory and practice always mix. Thus, every technician is a bit of an engineer, and vice versa. And certainly every thinking human being is a philosopher, regardless of whether he philosophizes well or not, knowingly or not.

What rational philosophy has to offer to engineers and technicians is much more friendly, speaking not like a boss but like a friendly neighbor who may offer both criticisms and alternative proposals but not commands. And the chief criticism philosophy has to offer to modern engineers and technicians is of their organizational method: it is too hierarchical and so invites unnecessary bossing. The alternatives it proposes are diverse, but chief among them is that we all develop a friendly attitude towards our critics and towards the people we criticize. This may be translated into many detailed proposals.

Here are important examples. First, most people mix small-talk and shop-talk with sales-talk. This is terrible. Try and hear yourself, and make an experiment of reducing (and perhaps even eliminating) your sales-talk and see if you do not like yourself better this way. One way sales-talk pollutes is that people bend their views to suit their sales-pitches. For example, in a buoyant mood engineers boast that every experiment is repeatable to any desirable practical degree of precision, but when in a defensive mood they declare that no experiment is assured of ever being properly repeated. This kind of inconsistency makes poor sense and presents people in a poor light. Second, it is common to view it as responsible for every person in a responsible position to stick to his specialty. Breaking away from this rule sounds as either the claim for expertise in many fields, or as irresponsible, or both. This is particularly important in view of the fact that most experts are trained in physical technology. There are as yet too few trained social technologists around, and some of these are poor specimens indeed. A responsible person evidently confesses ignorance while raising problems that bother him – especially when the problem pertains to his job but not to his specialty, as occurs all too often in real life.

The major concern of rational philosophy should perhaps remain with individual human beings: only individuals suffer and have problems and aspirations and hopes – even when they suffer in groups and as groups. Yet when they suffer as groups, those who wish to be helpful to them have to address them as groups, whether as nations, as oppressed minorities, or as the under-privileged. And technology is almost always a group affair and has its social aspects; especially the problematic ones are sadly neglected. Many simple improvements in the public domain which can be implemented are not implemented unless and until disaster strikes. Until we know why, this will remain the norm. Consider safety in case of fire. The cheapest and most obvious improvement is to make doors of fire exits open to the outside, rather than to the inside, which is particularly important for places of large concentrations of people, such as public theaters where fire may alarm people to crowd against doors which, if opened only to the inside, may stay closed persistently just when they should open most urgently. This obvious fact was discovered in the dreadful empirical way rather than thoughtfully as it should have been. Is it possible that no one thought about the matter? Clearly, it is more likely that someone thought about it yet was not listened to until disaster struck. The movie *Jaws* that was seen by tens of millions of people describes an even greater irresponsibility – of town fathers of a seaside resort suppressing information about sharks for obvious financial reasons. How true are such stories? We simply do not know as yet. Airline pilots have a saying: no airport is improved, they say, before blood is spilled on it. This matter should be examined, and, if true, it should call for political action.

Without placing the philosopher above the engineer, then, we may notice that problems of control are present everywhere, including the problem of controlling technology in general and the checking of the situation regarding controls in general – checking motivated not by pro-technology or by anti-technology propaganda, but simply by the desire to find out how things stand and how best they may be improved upon.

Without further ado we can examine the general answer endorsed here. Whereas problems regarding specific controls belong to specific technologies and are solved best by students of these technologies, the general – philosophical – problems of the control of technology in general are not best handled by philosopher-kings. Rather, they invite the development of appropriate means of democratic control. If people will wish to try and implement the suggestions and proposals made in this book, then they will have to act as citizens of a democratic state, not as citizens who have license to rule or to act in public simply by virtue of their (true or alleged) superior knowledge. The pretext that the present state of emergency calls for the suspension of democratic control is seriously defective, as the urgent need is exactly for the implementation of new means of control on all levels – local, national and global – and these can evolve only democratically.

2. DEMOCRATIC CONTROL IN GENERAL

Learning is traditionally divided into specific fields of study or subject matters. These evolved organically during their prehistory and early history. Fields or subject matters cannot be easily characterized (and certainly not by the use of the traditional means, namely the Socratic-Aristotelian theory of essential definition). Yet this much can be said: the idea of a field or a subject matter evolves, like most ideas, intuitively; and it crystalizes, like most ideas, through critical assessments and canonic formulations of theories. Thus, fields of study are traditional social institutions.

The function of dividing inquiry into fields or into subject matters is two-fold: to facilitate and to limit study; or to facilitate through limitation; or to limit while facilitating and thereby opening new channels of thought. That limiting may be accepted for the purpose of facilitating study and research is obvious. One ready example is the tacit agreement of physicists to overlook the problems caused by animal motion, so-called: they had enough on their hands studying the movements of billiard balls even if they altogether ignored the human behind them. In the social sciences, most imposed limitations are very harmful and may at times be parts of some overall ostrich policy imposed by irresponsible governments. Ostrich policies need not be imposed: at times they are chosen freely. What ever our conception of technology is today, it includes not only physical engineering, but also the study of the spread of its services among nations. The fact that the Western world depends on imported fuel and raw materials is obviously a political matter of great technological concern. The result that such a fuel-exporting country as Saudi Arabia is one of the richest countries in the world when gross national product divided by the number of inhabitants is the measure of wealth, but one of the poorest where the welfare of humans is concerned, where traffic in slaves and no rights for women or for aliens is concerned – this result is obviously political. And by a policy chosen by most Western technologists, all this is declared outside the field of technology. The political concerns that emerge very quickly and obviously from any general study of technology are usually such that are characteristic of the industrialized world; they have to do mainly with the ecology of Western countries or of Spaceship Earth as a whole. The ecological impact of technology on poorer countries, if studied, concentrates on overpopulation and the poverty which results. Yet the fact that some of the poorest countries in the world squander their fortunes on luxuries for their ruling classes and on the most sophisticated weaponry systems – this fact is left dangerously outside the specialized concern of technology and the policies which direct it.

When technological innovations are entered into the social system, with a resultant change of the system, this inevitably raises new problems within it. The disposition of technologists is naturally to tackle these new problems again by the application of more technology. In order to make this process more rational, we simply must broaden our concept of technology to include social and political technology. If social technology in the coming decades will be nearly as successful as physical technology was, our globe will soon resemble what early dreamers deemed a mere utopia by comparison. Indeed, the obvious proposal is to channel our limited resources devoted to research and innovation differently: less to physical and biological technology and more to social and political technology, especially regarding democratic control and the training for democracy. There are two factors preventing this, and they reinforce each other. First, unlike physical and biological control, social and political control may involve the manipulation of human beings and the setting of constraints on their freedom; and social and political technology necessarily involves social and political control. Second, social and political technology and control involve the world of politics and other social activities not given to traditional rational control since politicians and social activists often represent conflicting views and conflicting interests, and there is no theory as yet of rational dissent and of rational compromise.

It is the ever growing credibility gap which is the most serious threat to our political life, and it is political irresponsibility that grows behind this gap, and the disregard for population explosion, pollution and nuclear war. Hence there is no way of controlling our ecosystem without regaining political credibility. Hence, the most urgent task of the philosophy of technology, of the concerned technologists, of the ecologically minded citizen (who still maintains some sense of proportion), of the responsible citizens of their countries and of the world, is to contribute to the study of the present, appallingly low, level of political credibility and to the attempt to raise it to a manageable level. It is clear that under conditions of great emergency, when lives of millions are concerned, responsible individuals are under enormous pressure to deviate from the moral rules learned in stable society to fit and foster stability. Much that looks atrocious under normal conditions appears imperative under enormous strain. It looks as if to torture a captive and then deny the fact is less of a deviation from ordinary morality than risking the lives of some young soldiers. And so the temptation to lie is enormous when the choice is between lying and a military operation, for example. Yet this is a fatal error, as it is the crack that widens into a credibility gap.

From year to year the risk of a global ecological disaster is growing, yet nothing or almost nothing is done about it. The reason seems obvious. People who become sufficiently aware of the centrality and gravity of the problems of global magnitude tend to devote their time, at least their free time, to global problems. Nowadays these people are mainly ecologists.

Under the strain of the sense of urgency of the problem they feel they must advertise their findings and the urgency of the problem. And then they naturally tend to exaggerate. Thus, in addition to the rapidly declining credibility of politicians we now witness the even faster declining credibility of ecologists. Rational communication is thus endangered and with it the hopes of survival: we are sinking into helplessness.

And so it may be apt to conclude this introduction by a call for truthfulness and for the search of the very establishment of social and political norms that should put truthfulness as the very highest requirement on the ground that losing it is losing all. Truthfulness, however, does not preclude error, only willful exaggeration, not to mention lies and concealment and self-deceptions and, worst of all, half-self-deceptions. Anyone impatient with the tenor of this book, with its leisurely and sedate pace, may remember that it is dictated by a sense of urgency distilled out of utter despair. If one finds this agreeable, one may wish to read it critically and try to improve upon its mistakes – again with leisure and due care. We may have no time for that, but it is the quickest way: the tortoise always wins the race with the rabbit, and the rabbit is too frightened to stop and think why he keeps losing. Time is short; very short. We must now stop and think. We must reform our most basic presuppositions, and this requires that we stop and think. Since time is very short, we may have to start developing some techniques and truthful critical thoughts straight away or lose all hope in learning to control ourselves as urgently required – not only as individuals and nations, but also as a species.

PART ONE

TECHNOLOGY AND CULTURE

INTRODUCTION TO PART ONE

We need a proper setting for technology, a setting which may serve as background for a more integrated picture of technology then was hitherto offered. Yet it would be an excess to seek an ideal setting. Instead, what will be attempted here is the elimination of two popular errors which may be viewed as parts of current popular mythology, so pervasive and pernicious they are. The one is that scientific technology is more akin to science (it is usually part-and-parcel of science proper) than akin to art; the other is that technology calls for elitist government (indeed for technocracy). In the following pages technology, no matter how scientific, will be presented as art, no less than as science. The case of architecture illustrates how trivial this view is; but this view raises important questions which are poorly answered, thus complicating a simple matter. And the rise of technology to its enormous present power and (just) prestige will be presented as inviting more advanced means of democratic control, on the democratic principle that control raises credibility.

CHAPTER 1

TECHNOLOGY AND LEARNING

It is customary to begin a study with a definition of the field of study. In the
present case what is needed is not so much the definition of the philosophy
of technology, as the definition of technology. Technology, however, can be
defined in different ways, and the choice of definition, which may seem
rather innocuous an affair, may prejudice our attitude towards it. Different
definitions and/or different circumstances may lead to different prejudices.
So perhaps we are better advised to move backward – not from the choice
of definition to the imposition of a prejudice but from a choice of a prejudice
to a definition that backs the prejudice. The prejudice here is the idea that
human survival is our responsibility, that it can best be democratically
sustained, and that democracy invites the development of better techniques
of government and control, namely better techniques of increased public
participation in government and in control. Technological thinking has thus
far led to the rise of anti-democratic technocracy; it can be further developed
to help create the democratic antidote to technocracy. Technology may,
then, be defined – democratically – as any means of social control (of
anything).

1. TECHNOLOGY AND SCIENCE

The proper place to start the study of the philosophy of technology is to find
the place which thinking about technology occupies in the world of learning.
Quite generally and with as little finesse as possible, we may observe that
the most general division of the world of learning is into the arts and the
sciences, with technology in the division of science as a matter of course.
The sharp and clear description of the division can be found already in
H. G. Wells' science fiction *Things to Come*, where, in the grand finale, the
scientists of the distant future launch a rocket to the moon despite enormous
opposition from the world of art. Wells is even more savage in his earlier
Time Machine, where he describes mankind as split into two species, the
softies who have evolved artistic sensibility into a way of life but now have
lost all science and the hardies who are so mechanically inclined that they
tend machines even for fun but have lost all sense of beauty; and with this,

11

we are led to understand, they have lost science and even the advantages of technology. Today's division of learning into the arts and the sciences is very simple-minded. It puts on one side history, together with language and literature and the fine arts and music; on the other side we have physics and engineering and the life sciences and medicine. When C. P. Snow, Lord Snow, delivered his famous lecture, *The Two Cultures*, he explicitly omitted the social sciences and described his concern with the estrangement between the arts and the sciences, and he described the arts as aesthetical, sensitive, intuitive, soft, unlike the sciences which he described as hard, mathematical, factual, employing cold logic. This enabled him to view the social sciences as part soft – art – and part hard – social statistics and all that. Nevertheless he spoke of an unbridged cleavage. For examples of hard science he took the second law of thermodynamics and machine tools; one is an article of theoretical physics and the other is as technological as possible. The two seemed to him to be one field.

This is a dangerous myth. One could attack it on various lines. One could claim that creativity is not limited to the arts, that mathematics and science depend on it no less than the arts. One could claim that discipline and techniques are not limited to the arts, that mathematics and science depend on it no less than the arts. One could claim that discipline and techniques are not limited to the sciences, that the arts and the fine arts depend on them no less than science. That is to say, both the sciences and the arts require both techniques and creative intuition. (It is tempting to interject here the obvious and regularly overlooked fact that true technicians are as soft about their engines and as imaginative about them as artists are about their own materials. But let us stay in generalities.) One could also divide all thinking into the contemplative and the practically oriented, so as to have both arts and sciences, both pure and applied, with technology and techniques on the applied side: techniques to cater both for bare essentials, like food and shelter, as well as for luxuries, like gracious living, including such techniques as wine making, which involves all sorts of activity, from the fertilizing of vineyards to the cultivating of good taste. These then are the obvious two lines of attack, the one showing both soft intuition and hard discipline necessary both for intellectual pursuits and for practical pursuits; the other showing both to be common to both the arts and the sciences, and that technology proper combines both. The attack on either of these two lines seems both very easy and very deadly. Yet neither attacks will be convincing: the myth seems very sturdy.

The reason for this sturdiness of the myth of the division between art and science and of technology being on the side of science lies elsewhere – in a domain untouched by the two lines of attack described in the previous paragraph. The myth rests on the fact that makes Western civilization so very different from any other. Obviously, what makes Western civilization stand out more than any other is Western technology – the means of mass transportation and mass media, big machines and small gadgets, sophisticated surgery and wonder drugs and all that. These are all science-based.

There are, however, things that we do not want to be replaced by gadgets. The artificial heart is partly here partly in the stage of planning and experimentation, yet we all prefer the original to its substitute whenever possible. And no matter how magnificent the Western style muscle substitutes are – the bulldozer, the power engine, the powerhouse – we want to retain our muscles, and try out the Canadian Air Force exercises and jog and frequent Yoga classes. Yoga is a typically non-Western technology. Can we do without it or its substitutes? Should it not count as technology?

This question – is Yoga technology? – does not differ much from the lines of attack described above as deadly yet unconvincing. There is no fundamental difference between the fact that architecture is both art and science in the vulgar sense, the fact that piano playing includes both music and five-finger exercises, and that Yoga is a technique to maintain good body condition and peace of mind as well as a refined aesthetic and religion and more. Yet, whereas in Western terms technology is at least at first glance science-based, in India it is so steeped in religion and so indigenous as to defy Western categories. Now, we can say, never mind India: the modern Indian society is acquiring paraphenalia of Western civilization faster than the West is acquiring Indian influences of any kind. Yet here lies the strong reason for the sturdiness of the myth of technology as a part of science, and it can be put thus.

The West influences India more than India influences the West. Yet this is no proof of superiority. India must learn from the West in order to survive, and so the resultant loss due to Western influences may be unavoidable. Perhaps the loss of spirituality in the West is disastrous in the long run and may be regained only by the encouragement and enhancement of Oriental influences. This argument is very appealing and seems to support the myth in question.

This argument exhibits much good will and quite possibly uses correct information. But it does not support the view it is meant to support. There are two views about Western civilization which are similar and thus taken as one. There is the view that the West is overspecialized and that its diverse specializations fell apart and want reunification. And there is the view that the world is overspecialized, with the West having developed technology, become materialistic and lost its spirituality, whereas the East, etc. Certainly we all want the arts and the sciences in the West to come closer to each other, and we all want the West and the East to come closer to each other. But over-praising the East so as to compensate for its technological backwardness will not help anyone. On the contrary, we cannot avoid noticing that unification invites either uniformity and thus conformity and thus oppression, or pluralism. If we want pluralism we must allow diversity without constant fear that parts will envy other parts and feel inferior and simply demand compensation – much less compensation by cheap means.

It remains incontestable, though, that we want a new civilization which will be both diversified and global, a multitude and a unity, and with the whole of mankind as one entity, socially as well as politically, yet without losing its pluralism.

The most famous defense of pluralism is a philosophy known as pragmatism and often identified as American though its native soil is German. Pragmatism is, first and foremost, a theory of truth: it says that a true belief is any belief worth holding. It is a preposterous theory, since in fact only the truth is truly worthy of belief. Yet this fact is problematic. If I hold the truth to be one, am I not putting my belief above yours? In other words, possibly the belief that my belief is true and that truth is one may make me intolerant of your view and thus it is a belief which makes its holders dogmatic. And, admittedly, pragmatism was endorsed in order to allow for pluralism.

Technology is by definition a means, and so pragmatic. Many philosophers who view truth as very problematic wish to view science as devoid of truth and hope to achieve it by viewing science as identical with technology. This, of course, enhances the division of culture and civilization into art and science, since thereby science becomes the same as technology. But we can look at fine and applied art the same way and deny that there is such a thing as fine art since beauty is no less problematic than truth, and then both the arts and the sciences will be parts of technology. Technology will encompass then the whole of our culture. Religion or history then will not be merely a possible means of government propaganda; it will be nothing but government propaganda! The government's propaganda will then be necessarily true for the nation and hence whatever its critics from within way will be sacrilege and necessarily false! This is a nightmare that can result from taking technology, which is a means, to be its own end. The myth of the Golem or of the Frankenstein monster, i.e., of heartless science taking over, is not possible as long as science is viewed as the search for the truth. In such a situation, technology may – but need not – get out of hand. But giving up the idea of truth, of the one and only truth, renders technology the same as science, and thus it must get out of hand.

The pragmatist attacks the opposite view, the view of truth as one, as dogmatic and intolerant. This attack was answered before pragmatism gained its popularity – indeed, the answer itself led to the rise of pragmatism. Two hundred years ago the philosopher, publicist, and playwright Gotthold Ephraim Lessing answered the charge by a fable: A father gave his three sons three rings – one of them genuine, two fake – and none knew which was which. The three rings stood for the three Western religions; the reason Lessing used a play and a fable is that it is not at all clear whether he held scientific truth to be on a par with religious truth. Those who say truth is one, usually add, and science (probably or certainly) has it. Those who say, we have plurality in science just as in religion (or else we could have no case of Newton being supplemented by Einstein), usually also say that pragmatic truth holds both in religion and in science. To break from these narrow confines, let us observe that we cannot compare truth claims of different religions though we can compare those of competing scientific theories. So truth is one, but for science it is but an ideal: one scientific theory replaces another when it comes closer to the truth.

This view need not be dogmatic, yet it is not too pluralistic. It is quite all right that all physicists prefer modern theories to the classical ones, but this does not mean that all physicists agree on all matters physical. Moreover, there is more disagreement outside science than within it. How is rational disagreement at all possible?

A brief glance at the present fundamental presupposition then: truth is one, but only as an ideal; at times it is easy to apply this ideal to existing disagreements, but most often even its application is debatable – some disagreements are eminently rational, some are eminently not, most of them are in between. In particular, a debate may start as rational, but one side may lose more and more conspicuously; staying with it becomes increasingly irrational. Now, traditional views of technology were eminently rational; they are becoming increasingly irrational to the point that they endanger life on earth. This is surprising, since technical matters offer less room for rational debates of fundamental and abstract matters: techniques are easily tested by implementation so that claims for truths about them are easily decidable, and so there seems to be no need for a unified theory of techniques for their successful application, so that technology has seemingly no need for abstraction. Yet this very fact has finally created a crisis: we want diverse techniques and the freedom to apply them in diverse ways, yet we want an overview and means to control all techniques so as to avoid global disaster. We likewise strive both for technological progress and for international exchange, yet we want social and physical technology integrated. We want both diversity and the striving towards common goals, including truth and world safety.

The conflict can be eased by noticing that technology allows for more diversity than science, since science is cumulative in a sense, yet technology is not.

2. TECHNOLOGY AS NON-CUMULATIVE

Ever since *Robinson Crusoe*, the dependence of sheer survival on ingenuity was a standard topic of fiction and of discussions of diverse sorts. Ever since Jules Verne's *Mysterious Island*, the value of science for survival enhanced the picture. The heroes of that novel make a lens to make fire, use trigonometry to estimate the height of a cliff, and so on. On the very popular television series *Gilligan's Island*, the scientist on the island produces electricity. The truth, regrettably perhaps, is nearer to E. M. Forster's classic *The Machine Stops* of 1909, which describes a society disintegrating with its loss of technology. Even *The Admirable Crighton* saves his marooned company not by scientific knowledge but by his leadership and practical skill. All this fiction of the desert island genre offers much food for thought.

Science is not cumulative in the sense in which most people still believe: science is not a house built brick by solid brick; in time each contribution to science is modified, if not rejected. After Newton, Copernicanism altered radically; Einstein radically modified classical physics and placed it in a surprising new context. Nevertheless, science is cumulative in a more flexible sense: the contributions to science are stretched, contracted, remolded and change color. Enormous amounts of informative details about electricity and magnetism to be found in books two hundred years old are omitted from their modern substitutes. Yet they are easily retrieved in improved form: they are hardly lost.

To the extent that scientific information is technologically applicable with the aid of common sense, the technology thus accrued is also cumulative. Science fiction authors take liberty with this fact. For instance, in the popular *A Canticle for Leibowitz*, the rediscovery of electromagnetism after the future atomic holocaust and the ensuing second middle ages enables those who comprehend them to design electromagnets and dynamos without further ado. In historical fact there was a long hard transition from electro-dynamics to today's dynamos, or even to Edison's dynamos with which he energized the lightbulbs all over New York City.

In this sense technology is non-cumulative. We risk losing the ability to operate ocean-going sailboats because of the advent of the motor-powered boats, as is well-known: in small part we counteract this risk by the con-tinued maintenance of some sailboats by sea scouts, by sports clubs, and by the navy. This is not to say that the art is totally preserved. Of course, it cannot be, since ocean-going sailboats belonged to an organized inter-national network, subject to international law and custom and linked to sailors' folklore, not easily forgotten by readers of Joseph Conrad.

The British historian of technology Donald Cardwell ended his report of how much effort and learning went into his reconstruction of an early – Newcomen – steam engine saying: "Despite my Ph.D. in physics and subsequent practical experience I would only just be qualified to be an engineman in 1712" (G. Bugliarello and D. B. Doner, editors, *The History and Philosophy of Technology*, 1979, p. 9). Of course, this is not because an engineman two hundred years ago knew more science than a professor of physics today, but because much of the detailed useful knowledge available to the engineman in question by tradition has been entirely forgotten. The same holds for singing a Mozart piece, as Karl Popper tells us in his "Towards a Rational Theory of Tradition" (*The Rationalist Annual*, 1945; also his *Conjectures and Refutations*). (It is to some extent even true of science, of course, as can be seen from the effort that reading Archimedes' book on floating bodies requires; but the phenomenon is much less pronounced in science than in technology since by and large science is cumulative whereas techniques – of making a Newcomen engine or of singing Mozart – are not.)

One should not romanticize about lost arts. They are usually replaced by better arts or otherwise made redundant. The art of mummifying, perfected by the ancient Egyptians, is lost, yet certainly nowadays it is very easy to mummify with newer and better techniques. The arts that belong to lost fashions are forgotten largely because we outgrew silly ideas about the attractive appearance of the distorted and scarred human body, especially the face (including the tradition of fencing, in German universities, which produced scarred faces). Forgetting these is largely to the good.

Yet some lost techniques have to be rediscovered in appropriate moments,

such as the use of local plants as anti-toxins, known to the primitive locals
better than to the invading scientists. And, of course, trite as these details
may generally be, or vital as they may be in moments of stress, they prove
the point: a technology may be lost due to progress, a new technology which
replaces an old one may make the old one forgotten. It is hard to say how
many expert metallurgists could build any primitive furnace and find a metal
ore in the surrounding area and forge a primitive metal instrument. Of
course, we all expect the expert metallurgist to be better at it than the less
fortunately equipped. This is so because technology involves scientific
knowledge, we remember, some of which is applicable with the mere use of
common sense, some of which is applicable with some level of ingenuity but
not much. These will help the modern metallurgist build a primitive furnace
from scratch, no doubt, so that he will be better at it than modern non-metal-
lurgists, but probably not as good as the original primitive metallurgists.
Probably an utterly ignorant member of the civilized world is much more
likely to build a primitive metal furnace from scratch, say, on a desert island
than a person who never came in contact with a civilization able to use metal.
Yet this is not to say that a modern metallurgist knows all he needs to know
in order to construct a primitive furnace. How much he needs is a question
of fact, not a question theoretically decidable; in the parallel case of scientific
knowledge proper, however, the question is theoretically decidable, and even
decidable with relative ease. The difference between science and technology
here is a matter of principle.

There are two major problems involved in all or almost all transition from
sophisticated technology back to primitive technology, one individual and
one social. The individual problem is that of personal readaptation to
hardships; it was described in Verne's *School for Robinsons*, in Jack London's *Sea Wolf* and in Forster's *The Machine Stops*. The social problem is
that of overcoming the disuse for the social system and organization that
supports the specialized sophisticated technology and the need to develop
anew a social system adequate to the more primitive technology.

When the transition is made from a sophisticated to a primitive technology, unless it is done on a strictly individual basis, it involves complex social
organization and reorganization. This fact has hardly been studied, beyond
fantastic stories of how the collapse of a large-scale sophisticated technology
causes chaos and how individuals ("Lot") have to struggle against the chaos
while recreating primitive technology on an individual basis or on the basis
of the most primitive social organization – the nuclear family.

There is much truth in all this, yet it is really only a minor part of the story.

It is understandable that contemporary concern goes in the direction of facing the disintegration of large-scale sophisticated systems, but the concern does not go far enough, as a thought-experiment may illustrate.

Suppose an atomic war starts without causing the expected immediate atomic holocaust – suppose only a few cities were attacked the world over and nothing else happens on the military or diplomatic side of things for a day or two. This will of necessity create an immediate and unavoidable stampede out of the major cities of the world before anyone would be able to discover what was going on. The stampede would itself cause enormous damage and loss of life, but the major disaster it will create will be the irreversible destruction of the social-technological complex that makes a modern city what it is: the main services – supplies of electricity, water, milk, bread – will all be destroyed at once and quite irretrievably. Even if the machines will be kept in the best shape, and raw materials, such as fuel, water, and grain, are still available (we can ignore the fact that cows are much more problematic), even then there will be the need to reorganize the social pattern that is essential to the services. This would prove to be an enormous task because after a stampede it is unlikely that a sufficiently large portion of the labor force essential to normal city life will return to work fast enough to make the damage reversible.

It is, to continue this line of thought, quite conceivable that an individual in the situation here envisaged would be able to struggle against the chaos which the stampede would cause, and that he would have to go it alone, with his nuclear family, at times also with a handful of people who will be close friends, neighbors and associates.

Nevertheless, clearly, there is more to it than that: there is at the one extreme the current sophisticated modern social technological organization, and there is at the other extreme the sheer chaos created by its enormously explosive disintegration. But all sorts of middle ground is conceivable, such as the society on an off-shore island near a city, which normally integrates with the urban conglomeration on the shore across the bay but under the explosive conditions just described remains relatively well isolated for the few days of the period of transition. There are many extant social experiments, even large-scale ones, to do with local catastrophes like large-scale earthquakes, or with local social and political collapses and regroupings, due to war, civil war, revolution, and even to do with wars in remote places, with the discovery in remote places of alternative resources which make local export industries collapse overnight, and many other interesting facts.

So much for the thought-experiment. It shows how little physical and social technology are integrated, and how much is left to be done in this respect. Since technology is non-cumulative, and since the social aspect or social organization of physical technology is an infant science, there is hardly a study of all this material, vital for survival though it surely is.

Yet, clearly, if survival under emergency conditions calls for the revival or the creation of relatively primitive physical and social technology, surely this must cause a radical change – and even an emergency – on the social side. But the social side need not be very primitive, and much depends on how much we can plan for such emergencies.[1]

For, it seems clear enough, if we want to prevent emergencies we should be prepared for them, and if we study – and train for – the social and political regrouping under emergencies, and if the venture is not doomed to be a total failure, then it will also function as a major deterrent, since the public will then know in detail what they would have to pay for some foolish rash moves of some politicians (*The Atomic Café*).

It is hard to say why technology, unlike science, is non-cumulative. To the extent that it is a bag of tricks, it is possible perhaps to systematize these to some extent and try to present a theory that makes some sense of them. The more sophisticated technologies offer examples to this effect, from the evolution of computer technologies into parts of computer science, to the development of parts of pharmacology, and in particular of toxicology, to biochemical scientific knowledge proper. Of course, in part, technologies are matters of training, and training can only be improved, transferred, or intensified, not ever eliminated. Also, of course, while some ad hoc techniques and procedures undergo systematization, others may be discovered, even as the outgrowth of the systematization, so that completeness is not ever to be expected. Some techniques are inevitably lost. Some can in part be recaptured by science and by later and more sophisticated techniques; yet even when recaptured, they are in a way lost. Once we realize this and observe that each technology is also bound to social conditions and organizations, then we shall be better able to capture techniques, describe them as parts of ways of life – as social anthropologists do with some primitive techniques – and so be better able to preserve them. The ability to so preserve older techniques is also a better way to ensure survival in case of catastrophe, and also to prevent avoidable catastrophe.

To conclude, a unified system of techniques is deficient unless it includes some details of the social setting in which the techniques are operative. Once we learn this, we can learn also, and with relative ease, how we can do away with fixed assembly-lines, with fixed working hours, with discrimination according to race and to sex and other lingering social ills. We shall then design better the social settings for all sorts of technological improvements and implement them even in poor countries and in the less fortunate parts of the rich countries. This aspect of technology – social technology – is hardly ever studied thus far, though it is now very popular among certain avant-garde students of the quality of working life engaged in change research – development economists and industrialists, as well as politicians and union leaders, sociologists, psychologists, public-health practitioners, industrial and computer engineers, who aim to design democratic worker-machine complexes.

Clearly, it is possible to attempt to unify or create theories about diverse techniques in their social settings and devise theories of control and their possible technical applications. Institutions for the control of some in-dustries already exist, but to date they are very much in their beginning stages. We may approach technology on a large scale and attempt to capture it and the modes of control it calls for.

3. THE TASK OF DEFINING TECHNOLOGY

The word "technology" is used loosely in different contexts and it is not at all clear how it may be understood in general. Scientific fields are traditional-ly defined by textbooks. The traditional definitions are static and hence inadequate: they ignore changes that textbooks undergo in time. To improve upon this, one may view a scientific field as the textbooks which belong to it plus the problems these textbooks give rise to, problems studied by researchers in this field, whose successful solutions are added to successive editions or variants of these texts. This improved definition is what is now called normal science, in the terminology proposed by Thomas S. Kuhn. But fields of scientific inquiry, he observed correctly, undergo revolutions, and after the revolution normal problems are forged anew. Also, of course, revolutions may create some new fields of science, or also destroy some fields of science. It is hard to say when a field of science is transformed and when it is replaced by another, but this is of little interest or significance.

When we come to technology the picture is radically different. Whole fields of techniques, studies and developments of techniques, may become obsolete and give way to new ones. Steamships and automobiles replace sailing ships and horse-drawn vehicles in a radical manner, so that we normally take it that sailing ship technology is as obsolete as horse-drawn transportation – despite the continued existence of sailboats and of horseback riding.

Transportation networks are transformed as a result of the switch from horse-drawn to motorized transportation; transportation administration and maintenance are transformed as well. The sailboats and the stagecoach are not transformed – they are replaced. The fact that at first the horseless carriage resembles a horse-drawn one had obvious technical reasons, as well as obvious social reasons: the horseless carriage engineers were concerned with engines and were glad to inherit the carriage as it was, and the public was used to the carriage and was glad, to begin with, to merely replace the horse with the motor. Yet by no stretch of the imagination can one speak of the automobile maintenance shop as a transformation, rather than as a replacement, of the stable. We see, then, that we intuitively judge the difference between modification and the replacement which goes beyond modification. Why? What is the classification scheme which tells us where to draw the line?

We can classify a technique or a set of techniques in different ways, all reasonable in their contexts or spheres of interest, but not in general. We can classify it by its aim, we can classify it as a set of means, and we can classify it as a set of techniques. As classified by its end, the carriage and the car – the highway and the bridge as well – are all parts of transportation networks. As classified as means, all motors are classified together as power engines. As classified by techniques, internal combustion engines differ from external combustion ones, chemical fossil fuel utilizing engines differ from nuclear energy utilizing ones, just as coal differs from liquid fuel, fission from fusion. Similarly, metal parts in diverse engines may come under one category, and then the metal blocks of engines utilizing diverse fuels will be classed together and as separate from engines built of ceramics or of other materials. Even flying machines of all sorts whose bodies are made of steel will be more akin to motorcars whose bodies are made of steel than to flying machines whose bodies are made on non-metallic materials or of light metals and alloys.

All this shows the diversities of ways we can and do classify technologies. (For more details, see Paul W. DeVore, *Technology: An Introduction*, 1980.) Some people are delighted by all this, others are disturbed. They may try to classify all technologies together as the making and maintenance of artefacts. There is no objection to this way of looking at technology, provided it does not obscure certain facts, yet certain very obvious facts are thereby often obscured: the purpose of using artefacts may be ignored when technology is so defined. When we consider that a primitive and an advanced technique can serve the same end, and that under some conditions the primitive technique may indeed be preferable, we may want to be reminded –

constantly reminded – of the availability of the more primitive techniques. To take a very significant example, the technique of using powerful magnets to extract shrapnels out of soldiers' wounds was replaced by far more powerful techniques of X-ray photography plus modern surgery. The primitive technology was forgotten very quickly, to be rediscovered as preferable under field conditions for relatively small and superficial shrapnel wounds. The same can be said of techniques involving no artefacts, such as the use of horses as power sources for transportation. Of course, even horseback riding usually involves artefacts, but not always. A technique based on the use of sophisticated artefacts, such as the control of heartbeats by the insertion of pacemakers in patients' chests, may very well give way to techniques achieving the very same end involving less sophisticated artefacts, such as taking medication; or which involve the use of no artefact at all, such as ingesting foods which contain the chemicals of the medication, or a life-style with plenty of exercise, or self-hypnosis, or the control of one's own brain waves. The preparation of the diet and the way of life may involve much sophistication, yet adopting the diet and the way of life may require little sophistication. The self-hypnosis and brain-wave control techniques go the other way: little sophistication goes into their design but their applications require much sophistication. Whole technologies, from agriculture to family planning, have evolved which are super-sophisticated, and whose chief and amazing and admirable qualities are put in their design: they are so designed as to require minimal sophistication from their user. The birth-control pill requires little user sophistication, yet this very sophistication is excessive for those who need it most, and – side effects apart – this makes the pill unsuitable for them. Sterilization is much more sophisticated than pill production but requires sophistication only from the ones who administer it, not from the ones who use it. Sterilization techniques, however, are defective in their very finality, unless reversal techniques are considered too.

control pill user is at times to control the size and growth of populations, and it is at other times the control of the size and growth of families, and it is finally also a woman's control of her own body. These three goals are extremely different, as anyone with fantasy can easily find out by describing logically possible models where one of these three goals is called for but not the other two, or two of them but not a third. And when speaking of populations, one significant factor in making the pill effectively usable is offering the population of young women the little education whose very absence makes the pill ineffective for them.

In brief, technology in general is not definable in any narrow clear-cut definition. As social anthropologists well know, no description of the techniques used in a primitive society can be adequate to any degree without an account of the values and structure of that society, at least to some extent. It should not surprise us, then, that with more sophisticated and large-scale societies this truism holds too – and even more forcefully.

This disturbs quite a number of people when they begin to think about it. They may be disturbed merely because they wish their intellectual system neat and ordered; yet they may think there is a serious problem here. They may feel that the serious problem is: in what respect is the modern Western or Westernized society superior to other societies? And, indeed, they usually have an answer ready: the superiority is due merely to better physical technology. If this were the case, the fact that some chemical additives to diet may be provided either as pills or as raw foodstuff, or the fact that self-hypnosis can serve as a cure, has to be viewed as a minor curiosity. We may then admit that, by and large, though not necessarily in all detail, mental control of the body is better achieved by the oriental expert, be he a Yogi or a Zen Buddhist; yet the physical control of the body through drugs, surgery, or otherwise is typically Western. The superiority of the East over the West in one respect is only one instance to show the unity of mankind: East can learn physical technology from West; West can learn Yoga from East. The readiness to classify some Yoga techniques together with some pacemaker techniques by their common end spoils this picture, and so seems perverse. Moreover, whereas we may agree with social anthropologists that practicing a technique is related to values and institutions, we may want to ignore these when we wish to devote attention to the study of sophisticated means – technology proper – and leaving values to users, thus highly widening the narrow choice of pairing means and ends, and thus, perhaps leaving Yoga out of the sphere of technology altogether as pertaining chiefly to values.

Now, the objections voiced in the previous paragraph are deeply rooted in the Western liberal tradition. It seems only highly commendable that we should develop the theory and practice of the control of our physical and biological environment, not of mental control, much less of the control of our neighbors. John Dewey was fully aware of it when he introduced the terms "social technology" and "social engineering". Nevertheless, the fact is that we do have social technology and social engineering, that we do control our own selves – individually as well as when considered a social body – and that we do control our neighbors to this or to that extent. If the liberal ideal of value-free technology and engineering were pursued to the fullest, and techniques of self-control and of the control of others were deemed useful to describe, then we would not object to describing technologies usable for evil ends. One need not hold the view that value-free technology is entirely possible, but if one is a liberal then one has to endorse the value of value-free technology to a large extent. It is easy to understand that liberals refuse to study and develop and perfect and describe techniques of, say, enslaving people (as sketched, say, in Aldous Huxley's *Brave New World*); it is likewise easy to understand the traditional liberal wish to stick to value-free technology; together these amount to the understandable traditional liberal proposal to exclude social technology from technology. But this proposal will not work. It is better to admit slave-making technology as a technology, especially when it involves sophisticated technical and scientific knowledge (such as the use of euphoric drugs, electric shocks and conditioning, mentioned in *Brave New World*). It is better to face the present problem of defining technology squarely than to dodge it by defining human technology as non-technology, and it is important to see that objecting to the knowledge of slave-making technology is classified by its purposes, and is therefore no different than objecting to the knowledge of prison building, of handcuff making, of whip making and of whipping.

The idea of a value-free technology is the root cause of our wish to study techniques and means, not ends; the end of value-free technology is the end of the liberal philosophy to which it belongs. Yet value-free technology, being value-free, can serve the ends of illiberal philosophies. There is no paradox here, since it is well-known that the liberal should tolerate the illiberal, not unconditionally but to a greater extent than the illiberal is usually ready to tolerate the liberal. The value-free attitude of the liberal must permit the classification of technologies by their ends. Classed by their ends, horse-drawn transportation networks and motorized ones serve the same purpose; classed by their maintenance techniques they are as different as the stable is from the mechanic's workshop; classed administratively, they are similar and different in other respects. Our intuitions, therefore, are quite pliant, as can be seen when comparing classifications of the same material in very different contexts.

At the very least, we must notice, then, the implementation of any technique whatsoever involves both physical and social activities. Of course, implementing an innovation of a physical nature, say, the horseless carriage, is rightly considered chiefly physical technology, despite its social aspects, such as the proliferation of gas stations all over the country, which must involve legislation. Also, of course, every act of social engineering, say legislating universal compulsory education, is social rather than physical, despite the physical aspects of implementation, including the proliferation of school buildings and of safe pedestrian crossings. Hence, even in the extreme cases technology must cross the interface between the physical and the social: pure cases do not exist.

This is a trite and obvious fact which has revolutionary corollaries, opening the road to far-reaching criticism of 18th-century classical liberal thinking, thinking which has traditionally gone into the making of modern Western technology. For, it was this thinking which invented the ideal of purely physical technology. Peter Caws has noticed that the word "technology" which is the descendant of the Greek word "techne" which may apply to any manner of doing any sort of thing, so that the Greek word "erotike-techne", which means ways or techniques of love making, should not really raise an eyebrow. Yet it puzzles us because we are used to the 18th century use of the word to denote physical technology alone.

4. EDUCATION FOR TECHNOLOGICAL SOCIETY

The classical liberal philosophers assumed that science is a central item in the curriculum of an adequate education, regardless of any other considerations concerning education, the curriculum, or professional training; they deemed it important that the average citizen should be reasonably well-educated and hence scientifically proficient. They took for granted both that scientific proficiency brings about technological proficiency and that a society whose average citizen is technologically proficient is a technological society. Conclusion: a reasonably acceptable program for public education brings about a satisfactory technological society.

This philosophy is long superseded. This fact calls for a rethinking, including the rethinking of the curriculum. Instead, educators have clung to bits and pieces of the superseded philosophy which may still seem acceptable. The main result is that some science is imposed on the average student causing much hatred of the subject. Two other results which concern us here are that in specialized science departments often science is taught as a handmaiden of engineering and that engineering is taught as devoid of all social aspects.

These three results are central for scientific education today. The resultant situation is of crucial importance in our social lives. It reinforces the obviously erroneous view that scientific and technological proficiency are the same. For, when the population is divided into the vast majority who have no idea of elementary mathematics – the calculus, matrix algebra, analytic geometry – and the rest, then the rest, the scientists, resemble each other simply because only they know elementary mathematics; yet elementary mathematics is elementary: the view that all scientists and all technologists are alike since they all know elementary mathematics is like the view that all musicians are alike since they can all read a page of music. The division of society into scientists and non-scientists while ignoring the political effects of science, is to invite technocracy or scientific elitism or some other meritocracy to replace Western democracy. The readiness to let things be is the acceptance of the risk to let democracy slip away in the vain hope that a technologically oriented meritocracy will be able to find the proper place of science and technology in our culture and in our environment. This readiness runs contrary to the democratic tradition which distrusts experts and tends to check their tendency to exaggerate the value of their specialty in our society. Unchecked, the technocrats may well put the whole of their society under technological control. Even if we assume that technocrats will not be hostile to the arts and will not be hostile to education in other than scientific and technological matters, even then the question will remain. If the current situation persists, how far will technology be legitimately extended? How far will physical and biological technology be developed and social and political technology neglected? Will such a trend not necessarily end in a technocracy which has no social and political guidelines? Will the technocrats of the near future then take it for granted that all technological advancements be available to the public? Will genetic technology be developed and made publicly available? Will techniques coveted by criminals be also publicly available? Will there be no social and political controls? Will this question be natural or social? If social science will be applied, then there will be social technology after all. Should, then, social technology be administered by technocrats or by a democratic system?

Alas, without being asked, these questions are gaining popular answers. Science and technology are these days already organized in order to go together. They share a common ethos. The words characterizing the new ethos, "hard", "tough-and-no-nonsense" and "formalize-it-or-dump-it", are not new; nor is the ethos itself, really, as an ideology; it is new as an ethos, as the spirit of the age, rather than of a school of thought or an intellectual leadership. "Tough" work has its rewards which are immediately visible. This, really, is the meaning of "tough" – hic Rhodos, hic salte! do not tell us how marvelously you did (or will) jump in Rhodos; jump here and now! Yet this is the contempt which the literate shows to the illiterate when he should rather offer to help him acquire an education. And so it is also the hardening of the intellectual arteries. This is what the present work is about: it aims to draw the limits of toughness: the toughness of the experts which rests on popular illiteracy is not tough enough!

The advocates of toughness will dismiss my strictures with a shrug, and they will argue in tough arguments, as follows. The dangers of toughness were noticed long ago. The economist Lord Lionel Robbins observed that in ignoring economics and centering on production processes we may build first rate factories which will not begin to produce because of the collapse of a market. Nowadays, of course, tough economics is extant and is supposed to take care of that. No longer do we allow for the surprising appearance of large shifts in markets which trade and industry cannot cope with. This can be generalized: whatever snags tough science and technology come up against, they may tackle them by tough measures. Tough measures are, and this is the point, tough because they are quickly tested for efficiency, and, if found wanting, get remedied or replaced. Hence, tough measures are deemed not dangerous. Hence, no concrete example of the dangers and limits of current toughness need convince us that toughness is risky or limited; on the contrary, it may serve as a challenge to broaden the applicability of toughness yet again. Moreover, the tough way to go about efficiently is precisely this: the places which should call for our attention first, call for quick and urgent improvement, of course. Hence, toughness is obviously the correct attitude! Conclusion: develop new tough specializations and open departments for them in the leading universities and educate young people to be tough in whatever they do.

The previous paragraph represents a social and political philosophy – including an educational philosophy – which is very important and well-known. It considers social affairs pragmatically – it is a social engineering, to use the term coined by the famous American pragmatist philosopher and educationist, John Dewey, and it is a piecemeal social engineering, to use the adjective appended to it by the most famous contemporary methodologist, Sir Karl Popper. (However, Popper advocates piecemeal social engineering on the condition that it is effected democratically.) The important point to notice is the genesis of tough (piecemeal) social engineering. We began with classical liberalism which recognizes only natural science as central for technological society and with science education for all. Science education then split to lay and professional, where the professional was the natural scientist and physical engineer. The need to control the free market somewhat has permitted minimal tough social engineering and professional social science education. But by now the emerging picture is all patchwork. It needs overall revision.

The major defect in contemporary education has to do with early specialization. The complaint usually voiced against specialized education is that it is limiting. This complaint is not very serious – it is diffuse and non-specific. All education is, of course, limited and insufficient: even the best educated person is willing and able to expand his or her horizon. The more serious complaint is that unless one is trained early enough and in specialized enough a manner in the hard sciences, one is never able to catch up with them. In the modern world, dominated by scientific technology, more and more people find greater parts of the modern world closed to them. With computer technology becoming increasingly pervasive, people lose contact even with the fields of their own specialization, and irretrievably so.

The tough-and-no-nonsense approach includes the myth that nothing can be done about this. A mathematician who has not "made it" by the time he is thirty – or twenty-five, or twenty – is lost, and the same goes for all mathematically rich specializations, for all hard fields. Similar myths hold for the performing arts.

Now these myths are false. Education can be remodelled in order to overcome the barrier to the comprehension of experts. And what is required is a total readjustment, beginning with the rejection of the hard-soft dichotomy, and with it the lay-professional dichotomy.

It is hard to believe that we can revolutionize education so as to make technological thinking available to all people with an average education. To attempt to allay his serious reader's doubt, an author may have to examine the details. The examination would make his book a treatise in technological public education, and so this task cannot be undertaken here. But a few hints will illustrate both the problems and the opportunities. The purpose of the illustration is to indicate that perhaps these – the problems and the opportunities – are misplaced. For, it is all too commonly agreed upon by educators, from kindergarten teachers to professors of medical education, that opportunities do dwell in the educational system and that problems stem from the limited capacities, attention spans and interests of the modern young brats. The facts go the other way.

Consider Maria Montessori, the educator who managed a minor revolution in her instituting the school system named after. her. She began her revolution when she noticed, with an enormous sense of profound surprise, that a child can concentrate intensely on a task for an enormously long span if she is interested in it. This discovery is still not generally utilized.

Nor has Montessori's proposal to utilise games been taken up as it should
have been. Computers are terrific tools and using them to calculate
probabilities will already make the simplest and commonest games of
chance, old and new, enormous sources of incentive for preschoolers for
arithmetic self-training. Language instructors are familiar with the fact that
a child's acquaintance – even her merely nodding acquaintance – with a
foreign language suffices to facilitate the acquisition of that language at any
later stage. The analogous fact is that some training with machinery in
childhood has similar effects: it destroys for good the all too popular and
pernicious inability to hold an instrument in hand or to try to locate a defect
in a simple apparatus. A child taught once to plane a table more or less well,
if he or she was taught the skill with no excess of disciplining and aggra-
vation, has already better training in manual skill than most people have
these days.

Many educators doubt the truth of the previous paragraph. This is to be
expected and encouraged. Yet no one will easily fail to notice that were the
content of the previous paragraph true, it would decidedly be of a great
significance. This will then naturally lead one to wonder, has the content of
the previous paragraph been examined? Why have the people in charge of
the educational system not gone Montessori? Of course, to some extent they
have. But to what extent and why? And have they examined the proposals
of thinkers more radical and more significant than Montessori, such as
Bertrand Russell and Albert Einstein? Where is the literature on that
subject? How is it that so many experts in education know so little about
education? All these questions indicate that all too often the educators in
charge of our future generations are merely not up to their tasks yet all the
same guilty of criminal negligence. Although many individuals are exempt
from this dreadful charge, no part of the complex Western educational
system is, neither the kindergarten system which forces children to play with
stupid sexist toys, nor the teachers' colleges, nor the professors in engineering and in medical schools and colleges; though, no doubt, the very worst
are the schools of education in the most celebrated institutions of higher
learning.

Those who are interested in details should read educational technologists,
such as Caleb Gattegno, to learn how problematic is the attempt to
implement the slightest educational innovation – especially those which raise
the technical proficiency of the average student, of the non-specialized
student even if he or she be brilliant. This holds for all attempts to raise the
technical proficiency of the non-expert, regardless of whether the proficiency

in question has to do with language, with writing, mathematics, the arts, or technology of any other sort. And the obstacles grow – they are not psychological but built into our social system which is geared to prefer the raising of expert proficiency to the raising of non-expert proficiency. This must be changed by accepting in practice the aim of making the language of the expert accessible to all.

5. CONCLUDING REMARKS

Scientific technology has invaded almost every part and every aspect of modern life. Yet the average, educated, non-technologically educated person knows least about technology and is frightened of both logarithms and the screwdriver. This is a terribly dangerous state of affairs, and should be altered by the reform of all education – from kindergarten education up. The commonly accepted mythology that justified the current state of affairs, or at least proves it unavoidable, is the identification of science and technology as "hard" and as "tough-and-no-nonsense", as well as the identification of the "hard" with the "tough-and-no-nonsense". Both these identifications are serious and damaging errors. These errors are now deeply entrenched in modern society since they are religious dogmas of the people who run the educational systems of the West.

Yet we cannot hide behind the defects of our educational system. The global disasters they prepare will hit us all alike. We must start a war on the evils of our educational system; we must run out of town the educator who accepts the technological ineptness of his charges and we must run out of town the educator who takes the technological aptitude of his charges to be a mark of exceptional distinction. Technology should be demystified by making its tools available to the general public at popular prices.

CHAPTER 2

TECHNOLOGY AND ART

Technology is often viewed as merely instrumental. Technology is also often viewed as having some aesthetic value too. When two conflicting views are repeatedly voiced, it is likely that something has gone amiss and should be attended to. The view that technology is merely instrumental is uttered in a certain context. Indeed, all disclaimers put in such terms are highly context-dependent since, obviously, anything can be viewed from different viewpoints. The view that technology is merely instrumental is usually a disclaimer not of its aesthetic value but of its epistemic value, i.e., of its value as knowledge. To be more precise, it is not the value of technology as knowledge that is disclaimed, but the technologist's need to explain or understand why his techniques are useful. Nevertheless, the same words often serve another disclaimer – to deny the beauty of technology. That disclaimer is obviously false, yet may be rescued when it is taken to mean that technology may be useful even when ugly, though it need not be ugly and may be beautiful. All this is rather trivial. It was presented here as a means to free the literature on the topic of the worthless misunderstandings which infest it.

1. ASSURANCE

The question, Is medicine|an art or a science, has been examined and toyed with for many generations. Even the meaning of the question has undergone change in various ways. In particular, most modern writers have overlooked the fact that the word "art" is Latin in origin and the word "technique" is Greek in origin, and the originals are synonyms. Thus, when we ask of medicine whether it is an art or a science, we may be asking, is medicine a technology or a science? Technology and science overlap but are not the same: characteristically there is science devoid of any practical application and perhaps applicability, and vice verse. There is always a part of technology that is ad hoc or governed by rules of thumb, a part that is non-cumulative, not science-based, and of little scientific concern. Thus, contemplating medicine as an art or as a science, with "art" meaning techniques, we can conclude that there is a medical art and a medical science. The same holds for all sorts of technology, from architecture to plumbing, from automobile design to automobile maintenance and repair, from oil painting to oil-picture reconstruction.

Inarticulate as the discussion about whether medicine is an art or a science is, it is clear that there is more to it. In particular, one may notice, the claim that one's professional work is an art rather than a science, is rather humble a disclaimer, and therefore usually made apologetically. Thus, when an enormous sum of money is sunk into the improvement of the acoustics of a concert hall to no avail, the acoustic engineer is almost sure to say that acoustics is an art, not a science. If he is also honest, then he warns his customer beforehand that perhaps the investment will be a total loss. If the acoustic engineer claims that acoustics is a science before he undertakes the job, and admits it to be an art only when he fails, then his disclaimer cannot absolve him, and his claim to science is open to censure. In this sense the acoustic engineer may (but need not) be a pseudo-scientist. A well-organized society will offer him incentives to behave well.

This can be seen more sharply by a contrast: when a physician operates on a patient with symptoms of acute appendicitis, only to find himself having removed a healthy appendix, he appeals not to art but to science: it is scientifically attested that about one-tenth of the acute appendicitis symptoms are misleading, and doctors are better erring on the safe side. This is not to say, however, that all appendix surgery can be justified or that any surgeon who appeals to science to justify his error is absolved. This is why we need control over medical error. Nevertheless, the very fact that an erring physician may honestly hide behind science in some cases but only behind art in other cases is illustrative of the point: the claim that medicine, or acoustics, or any other technique is an art rather than a science, is often a disclaimer concerning guarantees: science guarantees, art does not, and when one offers a guarantee without justification then one is in the wrong as a pseudo-scientist.

How much can science guarantee and how far can art go unguaranteed?

Finally we have arrived. These are, indeed, the central problems of traditional philosophy. Indeed, philosophy of science traditionally raises the question, by what criterion do we distinguish a valid guarantee from an invalid one? This is the traditional problem of the demarcation of science.

Let us take this point slowly for a moment. Though it relates to science proper, not to art, since traditions have altered under the influence of Einstein, and mainly through the enormously powerful and influential writings of Sir Karl Popper.

The Einsteinian revolution has demolished all hope to find anything like a guarantee in science: if Newton's theory has been superseded, nothing can assure us that its successors will do better in this respect.

The previous sentence looks obviously true to some and obviously false to others. It is, therefore, a sign of a profound disagreement or of a misunderstanding (or both). The claim that Newton's theory has not been superseded is repeatedly backed by major scientists with the true observation that it is still being used by scientists and technologists of all sorts, from navigators to artificial satellite launchers. Now, whatever one may say, some old theories are still in use (Galileo's, Newton's), some not (Aristotle's, Ptolemy's, phlogistonism). Some superseded theories are thus technologically better off than others (Newton's theory is better off than Ptolemy's). Nevertheless, intellectually they all share the same fate: they are replaced by a better view of the world; each of them is in the position of a mere predecessor.

Popper's theory of demarcation of science puts Newton and Einstein on the side of science and Aristotle and Ptolemy, along with astrology and necromancy, on the other side. Whereas scientific theories are refutable, he says, in the sense that it is possible to test them and for all we know find some of the predictions based on them false, the others are so vague that there can be no predictions based on them that may be in conflict with experiment.

Popper's demarcation between science and pseudo-science is admirable, since it exposes certain cases of pseudo-science as views that cannot be empirically criticized. A theory can resist criticism either because it is reliable or because it is shifty. Popper, and before him William Whewell, stressed the significance of this point. As Popper shows, things can get tricky: it is possible to present ideas which will not be undermined by experiment if false, but may be backed by experiment if true. To take an example from technology, we may consider the theory that there is a chemical cure for syphilis, cancer, or any other disease, or for all diseases. Once a cure for syphilis is found, the idea is verified, but we may try six hundred chemicals, fail, and still not refute the idea. Indeed, Paul Ehrlich called his cure for syphilis six-o-six because it was his six-hundred-and-sixth trial. Moreover, an assistant was so used to failure he let it go unnoticed and the discovery was nearly missed. Each of the six-hundred-and-five hypotheses concerning some given chemical's ability to cure the given disease is distinctly refutable and was refuted, but not the blanket hypothesis.

This example also shows that Popper's criterion of demarcation is wanting: it is important to note that Ehrlich's belief was not empirically testable, but this is no ground to call it pseudo-science. Also, some eminently pseudo-scientific claims were refuted: for example Faraday refuted many spiritualist claims in very interesting experiments. But as our concern here is with the demarcation between science and art, we should not pursue this point. Popper's demarcation does not tell us the demarcation between art and science. It is an art to try and cure syphilis by chemical means as Paul Ehrlich did, yet a science to cure it by the use of his medicine or by the use of antibiotics. If a researcher like Ehrlich embarks on a research program he has no guarantee for success, whereas today a patient suffering from syphilis in an early stage can be more or less assured of a cure once he falls into good hands! Here, then, is the difficulty people have in their effort to comprehend Popper: in a sense no theory is assured (of finality and irreplaceability) and in a sense some theories are (concerning antibiotics, for example).

The idea that research is an art has been traditionally contradicted by most philosophers of science. Sir Francis Bacon spoke of the draftsman's craft that used to be an art, yet with the introduction of the compass and ruler became a science. In other words, the artistic draftsman is not guaranteed success, but the scientific one is. And, said Bacon, research used to be an art: its findings were not assured, sparse, accidental. Given a compass and a ruler, the researcher will make discoveries in abundance – "in streams and buckets", as Bacon has so beautifully put it. This ruler-and-compass of scientific discovery is the inductive method. The inductive method, says Karl Popper, is a science-making sausage machine: the input is factual data, the machine's handle gets cranked, the output is science! Just like that.

A science-making sausage-machine, said Popper, is impossible. Nowadays, most philosophers of science agree with Popper, either because of the authority of Einstein, who said so earlier, or because Popper has a strong supporting argument: many great scientists who had made great discoveries also spent many successive research years with no find. But a century and a half ago, William Whewell too said science-making machines are impossible – one needs intuition and luck for success in scientific research – and John Stuart Mill declared Whewell an intuitionist and a defeatist and thought and declared that he had refuted Whewell by describing the science-making machine, known as Mill's four canons for induction. These four canons are still taught in many universities. Of all of Mill's wonderful ideas, none is half as well-known as this one really foolish one. If he knew how to make science, why did he not produce at least one lovely law of nature for us to credit him with?

Back to our difficulty. Popper says scientists must take risks; to be found in error if and when they are in error is the risk they must take. He says, scientific research is an art. But why do we accept the assurances of bridge builders but not of acoustic engineers? From whence the assurance of science? This question is overlooked by Popper.

William Whewell's theory of science can be presented with the aid of Popper's terms; such a presentation is sharper and more concise than the original. A scientific theory, then, is an explanation of facts and of older theories. It is a hypothesis (1) discovered somehow, (2) tested properly – i.e., attempts were made to refute it by new experiments – and (3) it withstood the tests. A theory which withstood the test, said Whewell, and Popper disagrees, is assured of permanent success: it will never be refuted. A theory which withstood the test, said Whewell further, must be endorsed. Does Popper agree with this point or not? It is better to avoid this question altogether, since science has nothing to do with any endorsement whatsoever.

Science in the post-Einsteinian sense, or in Popper's sense – that is, the theories that scientists propound and examine for explanatory power and for testable true-or-false predictions – science in this sense has nothing to do with endorsement, guarantee, credibility, forecasts, etc. Not so in technology. Science in the sense of art versus science, science in the technologist's contrast between art and science, is just the matter of guarantee. Science as an intellectual activity knows no guarantees: even Newton was finally superseded. But science as engineering guarantees some projects, not all; and similarly scientific medicine guarantees some treatments to have success, other treatments to have no success, and still others it deems to have possible success and possible failure. What kind of a guarantee is this? Unless we clearly distinguish the two senses of "science", we will not manage to be clear on the matter at hand.

Philosophers of science speak of guarantees as confirmations of hypotheses, accrued by the repetition of instances in accord with it. For example, each white swan confirms a little bit the hypothesis that all swans are white. Whewell and Popper object and say that we do not seek instances, we do not go where we expect to find white swans; rather we try to refute, we seek a non-white swan, and a theory is confirmed when attempts to refute it fail. Whewell says, a confirmed theory is guaranteed to be true. Popper says, no matter how well-confirmed a theory is, there is no guarantee that it will hold true the next time around.

We have to agree that confirmations are outcomes of tests, not of finding instances. And we have to agree that in science there is no guarantee. But this is not the case in technology, where guarantees are required in many cases and in civilized countries they are required by the law of the land: no guarantee, no airworthiness license for a new design of an airplane; no guarantee, no license to dispense a new medication. Truth in advertising means just this: the guarantee that advertisements are true. Now, it is not at all clear what is a guarantee, nor what needs licensing. Acoustic engineers are permitted to design with no guarantee. Designing an elevator for a high-rise building without a guarantee is legally forbidden in most civilized countries, and a guarantee that is very hard to obtain. Nor is this all. We do not have unanimity about what calls for guarantees. In most countries driving a car or a tractor is licensed, in some not; flying a plane requires a license everywhere. Licenses are given after some assurances are found, some guarantees. Rachel Carson started the ecological movement by demanding that no industrial plant be licensed unless a guarantee is given that it would not pollute the environment (*The Silent Spring*). Was she right? Are such guarantees needed? How are they to be found? Legislators have to know before they can heed her advice: the legislative process invited discussion and decisions on procedures by which guarantees may be procured. Carson had no adequate proposal in that direction.

Some technology, then, is by law a science in a legally defined way: it has passed legally specified tests. Other technology is not. It is hard, at times impossible, to say what makes a technology science, what makes legislation decide matters one way or another, what makes the legislators' decision at times wise.

2. INTUITION

Intuition is hard to pin down. Let us return to the question, is technology an art or a science, and agree it is in part art and in part science. Surely it is also in part sheer accident: one may need some technique to perform a given task and see before one's eyes just the right pieces of a machine come together in just the right way to perform the required task. Even then one needs some intelligence to notice the relevance of one's observation to one's task. And however incredible it is that a machine will assemble itself to do the required job, and however easy it is to notice that this is what happens when it does happen before one's very eyes, cases have been recorded where the machine did happen to assemble and perform the required task, and yet the observer failed to notice the fact. Oh, at times the pieces just fall together;

at times they are wrongly assembled for one task and happen to have been rightly assembled for another. Or it may be a living being naturally performing a natural function that is just what technologists are trying to perform artificially. It does not really matter. In any case, we may ignore the accident and center on the intuition of the dicoverer or the inventor – regardless whether we are speaking of a mere observation or of a design and an observation combined.

We can see at once that intuition sits nicely between science and accident. Thus, the job of the inventor and the discoverer – in brief, the job of the researcher – is not a science, or else we would have a science-making sausage machine; and it is not sheer luck, or else it would not be a permanent job. Where exactly do we place intuition?

This question is baffling and is by no means limited to research. To take a technological example, let us consider the technician with the proverbial golden hands. Golden hands are, by definition, rare. Yet they occur in each technology. We have the researcher with golden hands, the surgeon with golden hands, the garage mechanic, the designer or draftsperson (Bacon's metaphor was understandably an exaggeration: compass and ruler do not solve all drafting problems). Golden hands are beautiful to watch; yet we do not know where to place them in the scheme of things. They perform what cannot generally be guaranteed; yet they can be guaranteed to perform – and even to perform beautifully. Nor is it merely a feel: there is much more to it than a feel when a surgeon, be he or she a heart surgeon or a brain surgeon, is the artist with the golden hands. Surely they are not allowed to take chances as our proverbial acoustic engineer, or as the proverbial plumber. The acoustic engineer or plumber may be brought to court and called to pay damages, yet when a brain surgeon is brought to court, more than damages may be at stake: surgeons are not allowed to experiment with patients' lives. Yet one who could only do what other brain surgeons do, will not be rightly reputed as having golden hands. How does one manage the trick? How are surgeons permitted to use their golden hands?

It is easier to learn what the surgeon with golden hands manages to do than how. It is permitted, we remember, to have a standard rate of ill success in surgery. The standard may vary according to patients' conditions as specified by law and custom. The surgeon with the golden hands exceeds the standard. The surgeon with golden hands takes cases other surgeons refuse to take without thereby exceeding the standard. This is no mean trick. How does one manage the trick?

One may have dexterous hands; one may have really good hands; one's knack may, indeed, be in one's hands. In this case there is no mystery. Yet this is seldom the case. Dexterous hands are very useful for a surgeon, yet generally a person whose hands are not very dexterous is not a surgeon anyway. So this is not it. What then?

The word "intuition" is but a name for the quality described, for doing better than the statistically expected, yet with no known cause, reason, method, or system. By definition, whatever can be explained is no intuition. What intuition brings forth that can be understood, belongs to intuition no longer. Thus, most mathematical theorems and even whole mathematical fields or subject matters are brought forth by sheer intuition, i.e., by the act of some extraordinary minds, yet understood by normal minds with the aid of normal methods. This, indeed, is also true of many arts or artistic ideas, created by extraordinary minds but generally understood and repeatedly emulated by many.

This does not close the matter. There is a world of a distance between something that only a mind of genius can achieve and something that any average person can. Golden hands belong in the middle. Also, in the middle there are partial methods.

What a partial method is, again, is hard to say but easy to illustrate. Let us take a problem easy for a human to solve but very hard for a computer. Almost every problem posed in a high-school-geometry textbook is such. For sure, we can formalize both the textbook and the problem and then the computer will solve the problem and print out the solution faster than a human can say one word. Yet formalizing geometry, or even formalizing any old problem in any old geometry textbook, is not easy and may call for the expert's help, maybe for the help of an expert programmer with golden hands. Until then, an average or above average high-school student can try her hand at it and succeed. Girls are intuitive, say sexist teachers, so they'd better choose geometry rather than algebra. In algebra there are methods everywhere, yet in geometry not fully so. We have a heuristic – the word means rules helping to find, just as naked Archimedes running in the streets shouting Eureka! declared he had found his solution – and the heuristic comprises half-articulated rules of thumb one can fiddle with. Fiddling is no method; but perhaps it helps. Yet, more often than not, fiddling with a good heuristic helps often enough to count as a method of sorts.

Chess is a standard example for all this, but we need not go into it here. Anyone familiar with chess knows the science and the art of it, as well as the middle ground between the two.

Nor is this all. Intuition was called upon to divide between acceptable and unacceptable guarantees, and so it became a messy business. It led many to claim that intuition is – properly used – quite unerring. For, if intuition judges the propriety of a guarantee but can itself err, then we need a criterion to judge the propriety of the intuition that judges the propriety of a guarantee. This has led some thinkers to ridicule all intuition on the ground that sometimes intuition errs. Thus, since in mathematics certainty must be guaranteed on each step and since intuition is fallible, many mathematicians declared mathematical intuition worthless. Jacques Hadamard (*The Psychology of Invention in the Mathematical Field*, 1948) objects that with no intuition mathematical proof ceases to make any sense. What of intuitive errors, then? Hadamard does not touch upon this question.

Mathematics needs no guarantee, said revolutionary Imre Lakatos (*Proofs and Refutation*, 1963–64), but rather proofs and their refutations. Therefore, we need not fear using fallible intuitions when inventing proofs: intuition can err and then be corrected: we can educate our intuitions.

With this Lakatos heralded a new era: we can educate our intuitions. Of course, the same can be said of partial methods: they are fallible and can be corrected. Moreover, they usually stem from the general principles or the intellectual framework or the metaphysics of the field to which they belong and then the general principles which give rise to defective methods may have to be corrected (J. Agassi, "The Nature of Scientific Problems and Their Roots in Metaphysics", 1964).

Yet it is one thing to say intuition and partial methods can be corrected, and quite another thing to dismiss all guarantees. Why mathematics has no guarantee, why mathematical research is adventurous, Lakatos has discussed in a most lovely and inspiring and revolutionary manner. But mathematical adventures are not for technology: if engineers could not trust their mathematics when designing a bridge, then they could not be called to task or jailed when bridges collapse. Engineers designing and executing faulty bridges do end up in jail, and they cannot hide behind Imre Lakatos.

3. REPETITION

We are slowly approaching art proper, that is to say the delicate object of art appreciation, the object of aesthetic quality. The introduction of aesthetic considerations proper into technology, not to mention aesthetic criteria, may be required for the discussion of assurance, intuition and uniqueness. In any case, little is said on the topic other than to acknowledge some obvious facts: all too often technology and its products are as ugly as sin; sometimes their beauty is divine; this holds not only for specific objects but also for whole traditions and ways of life. The beauty of the excitement of youths repairing a jalopy seems a sufficient argument against those aesthetes and connoisseurs who sneer at technology and turn to the latest digital dolby reproduction of the most exquisite performance ever of Vivaldi's concerto for a viola d'amore or Schubert's Arpeggione Sonata on an arpeggione or Robert Cogan's No Attacks of Metallic Organs for organ and prepared magnetic tape.

The reproducibility of a work of art is what turns it from an art to a science. This is a hard fact. It annoys and bothers all aesthetes and connoisseurs of fine art. The celebrated art historian Ernst H. Gombrich has said that no reproduction, however good, can ever replace a good original. He even refuses to view photography and phonography as art, much to the chagrin and objection of his keen admirer, the art historian and photographer Carl Chiarenza and contrary to the life-work of Glenn Gould. And so, the irony of the last sentence of the previous paragraph may make good sense to Chiarenza and Gould, but will be lost on Gombrich, who will disdain any reproduction of any Vivaldi or Schubert (of Robert Cogan he will not wish to hear anyway, let alone listen to his music), dolby or no dolby, but will rather go to hear it in a live concert. Gombrich is right to declare something lost in the reproduction. But, observe Chiarenza and Gould, something is thereby also gained – not only technologically, but artistically too.

Engineers too are annoyed at people like Gombrich. Sir Ernst Gombrich will be sufficiently critically minded (as he must, being a disciple and a friend of Sir Karl Popper) to accept the verdict of a test of the sound of a Stradivarius against a good modern violin, of a live organ as opposed to an electronic organ, of a quadraphonic dolby reproduction as against the original. These engineers will say, in every case we can have a reproduction good enough to fool the expert. But they overlook something important. The excitement of going to a concert and witnessing a live performance is decidedly not reproducible. Nor is the excitement of a Hollywood screening of a raw film, or even of a gala performance of a movie in New York or in London or Paris. As Ian Jarvie has emphasized, even of seeing a movie on a first run in a local theatre is significantly much more exciting an experience than watching its rerun in an arts playhouse, not to mention seeing it on the T.V. on the late late show.

Art is fresh; reproducing it makes it a science, repeatable, dead. Not valueless, but not quite art either. What case, then, do Gombrich's opponents have? What possible case? For, they do have a case and they even have won this round.

To get to the root of this we must go more slowly. A concert is not its dolby quadraphonic reproduction, and by the same token a concert is not a reproduction of the first performance of the items on its program. Professor Gombrich will be all too glad to be transported to Beethoven's Vienna to join the audience of the first performance of the Ninth Symphony, and were any means of such a transportation possible, he would surely use them. Were technology able to reproduce that performance, he would be glad to witness the reproduction of the first performance in lieu of witnessing the event itself. Similarly, Professor Gombrich was not present at the unveiling of the Mona Lisa, yet he prefers the original to a reproduction, even though it is in this sense only a replica; a replica of its old self as it was in the unveiling, and for all we know a rather inferior one by comparison. Proof: if a time machine were available, Professor Gombrich would love to witness the original unveiling.

By high standards, no art is art except insofar as it is irreproducible, like a tasty dish or an ancient Egyptian piece of sculpture, which was on exhibition at a funeral and then left in the dark recesses of a pyramid to be seen no longer until the awakening of the dead. Yet the tasty dish is a replica of the chef's standard replica of the idea of the legendary originator of the conception of the dish just tasted, and the ancient Egyptian statue comes from the workshop, etc. In brief, by some standards no event is reproducible, by other standards every event is.

Now metallurgy is unquestionably technology, usually considered a science rather than an art. This does not mean that any step in its evolution was science, rather than art: the knack was important and the product, as the leading historian of metallurgy, Cyril Stanley Smith, tells us, was usually a decoration rather than a useful artefact. It developed in workshops of artisans or artists. Processes of casting, alloying, and welding, he says, began with works of jewelry and sculpture. More than that: "There is almost nothing metallurgical prior to 1900 A.D. that began by someone having a theoretical idea and then applying it." Indeed, the idea of science as art crystallized is a running theme in Cyril Stanley Smith's exciting writings, especially when he speaks of the influence of oriental art – past and possible future – on occidental art and science. He seems to think that oriental and occidental processes of crystallizations make the real difference between the arts East and West, not the arts themselves.

This is of crucial importance both ot the philosophy of science – especially of the social sciences – and to the philosophy of technology. Endless confusions which must be cleared up are due to oversight of the fact that every one of us is in a sense a reproduction of Adam or Eve, and as such (scientifically) reproducible, yet in a sense every one of us is unique and irreproducible (and as such a veritable work of art). The discoverers of cloning have questioned this and declared that since in principle genetic "information" is reproducible and since we are described by "message" "written" on our DNA molecules in a genetic "code", each of us is reproducible. Information theory tells us the opposite. No message is assured of precise repetition, no matter how hard one tries (Shannon's theorem; the central hypothesis of information theory). And we know that the genetic "message" is not confined to the genes. And we know that the genetic "message" will be "deciphered" differently in different environmental conditions. Cloning is no method of reproducing identical twins – if conditions are near ideal, cloning can bring about at most only almost identical twins. No twins are utterly identical anyway, not even at birth.

Do we not get finicky here? How close to identity do we need to get in order to be allowed to overlook differences?

We have to ask Professor Gombrich, or any other connoisseur and aesthete, for that matter. For, the standard of reproducibility is determined by the answer to such a question once it is taken without bias and applied across the board.

Is puppy love reproducible? Is love? According to what has been said above, it depends on standards of reproducibility. In one way, yes, indeed, puppy-love is reproducible. In another way, never. The same holds for any artistic venture and for any scientific experiment. Yet there is a great difference between reproducing a scientific experiment and reproducing a work of art. The standards of repeatability of a scientific experiment are given in the description of that experiment. This, indeed, is why scientists need not wait for Gombrich's answer: it is the crucial characteristic of both scientific experiment and scientific technology at large that in them each event is clearly described as repeatable under certain well-specified conditions.

This is not the case with an original painting, a score of a symphony, a recipe for a dish: these do not tell us what is a good replica of the original. The question remains – when is a reproduction correct? At times such a question is pedantic, at times of considerable interest. Does Glenn Gould reproduce an original Bach or a new version of Bach? Is the standard modern Hamlet or the standard 19th century one closer to the real thing? Is the film Hamlet Shakespeare? These interesting questions are open. In science, things are clear-cut: the question is not difficult to answer, as to what is a repetition of Oersted's experiment or of Halley's observation of his comet. These are in principle repeatable under known conditions.

This rather obvious yet often overlooked point has consequences for art which are quite remarkable. The music played on a record is art and the record itself is science, one can now say. We can be more precise. How much of the recording technique is art and how much science? Connoisseurs are all too often loath to ask this question. Consequently, and all too often, the artistic quality of the recording itself, the quality that calls for the recording engineer with the golden hands, is often ignored by connoisseurs – unless they are artists also devoted to music which is electronically amplified or reproduced or produced. And so we witness a situation in which the high art of electronic recording is too often wasted on a record of poor-quality music, while a record of a top grade performance of a great masterpiece is a piece of run-of-the-mill workmanship.

Let us take it for granted, then, that repeatability is a matter of standards and specification, whereas art can complement science – as that part of the irreproducible residue which was produced by a golden hand. But let us now center on reproducibility for a while, since, regrettably, connoisseurs often and rather pompously both oppose science and (scientific) technology as reproducible and defend the uniqueness that is the very soul of art.

4. SOCIAL TECHNOLOGY

Social technology seems non-existent. The fact is that it exists but is prescientific to a large extent. Why? Because of the almost total absence of social science proper to apply to social technology in order to render it scientific. This answer is too sophisticated. The following answer will do: social technology as yet knows no repeatability and hence no guarantee.

Much of what social scientists write is unscientific, and nevertheless very welcome. To criticize an idea as unscientific is often not to criticize at all. At times a writer wrongly claims scientific status, as Engels did in his renowned *The Development of Socialism from Utopia to Science* and as Freud did in all his writings. These claims have aroused Popper's ire. He hardly ever explained why this matters so much. Of course, as everybody should know, popular Marxists and popular Freudians and all manner of soap-box propagandists and street-corner preachers often claim scientific status to what they say in order to claim credibility. It sounds as if Popper says that science rightly claims credibility: Newton and Einstein deserve it, Freud and Marx do not. This is purely academic, since one cannot possibly believe both

Newton and Einstein. It may be taken, then, that Popper grants the best science extant its claim for credibility, including Newton in the last century but not in this century, and including Einstein today. But why should Newton and Einstein ever command credence? Can I decide to believe a Newton or an Einstein? Does it matter? It does not. Faraday rejected Newton; Einstein rejected Einstein; Lorenz and Michelson and others stuck to Newton and refused to move over to Einstein. Yet they were all great scientists.

Credence is all too inconsequential. Believe what you believe, and leave open the question: Is your belief scientific or consistent with science? Yet there is such a thing as a scientific fact, and if the word "scientific" is too presumptuous, the word "repeatable" will do amply.

Most facts are not repeatable. Also, when we claim a fact to be repeatable we may be in error. Yet, the whole tradition of modern science, since the foundation of the Royal Society of London, endorses the rule of repeatability. A scientifically attested report of a fact is a report presented as repeatable and also reported repeatedly by at least two independent and acceptable eyewitnesses. To count as scientific, an observation must include two items. First, a statement of it in generalization, with no reference to specific place and time. Second, it must include at the very least two independent reports of its manifestation made by independent witnesses as having been observed at specific places and times. The scientific observation reports, thus, have three parts: the generalization that says under what observable conditions what observable consequences happen; the report of the conditions and the consequences having all been observed by one witness at one place at one time; and then independently by another independent witness.

Little is agreed upon as to the rules and methods of scientific research practices, but this rule of reproducibility of scientific observation is religiously observed within science. Yet, traditional philosophers of science, with the notable exception of Immanuel Kant and Karl Popper, ignore this fact and take as scientific any observed facts rather than only instances of an allegedly repeatable fact. In particular, most philosophers of science in the middle of the twentieth century speak with Carl G. Hempel and Rudolf Carnap of the claim that instances of a generalization like "All swans are white" confirm it and discuss the why and the wherefore and the degree of this confirmation. In fact, "All swans are white" is an observation report which claims a scientific status as long as at least two people report having observed white swans and that is that: when two people or more report a repeatable observation to the contrary, such as "All Australian swans are black" or "All Brazilian swans are black-necked", then these too receive a scientific status.

It is clear then that scientific status is not the same as truth. Hence, what is required is to decide which scientific observations are true. Also, usually (and as was first observed by Newton in his *Opticks*, it seems, and discussed in detail by Pierre Duhem) when we overturn a scientific observation – declare false one previously declared true – we want also its correction: e.g., "All swans are white", when rejected, is replaced by "All European swans are white". Similarly, "the sun rises in the east" is replaced by "the sun appears to rise in the east"; and Ohm's law of resistance is replaced by the Faraday–Maxwell law of impedance (so that when induction is minimal the impedance of a current almost equals its resistance); and "the atomic weight of chlorine is 35.5" is replaced by "the average atomic weight of terrestrial chlorine is 35.5". And so on. All this leaves many problems open, but we need not discuss here any of them except what is relevant to technology.

Technology requires that certain factual claims, certain observable repeatable factual claims, be repeatedly tested – where tests are attempted refutations – and often along given specified lines. We are not allowed to begin our test of medication on humans, nor of planes' airworthiness. Only after tests on animals and in wind tunnels respectively were successfully performed and declared failures, only then are we allowed certain tests on humans, on volunteers of specific sorts, such as terminal patients and test pilots respectively. Often a terminal patient is willing to waive procedure, or a test pilot; yet in most civilized countries the laws and the underwriters involved forbid this.

So much for (scientific) technology in general, which usually means physical and biological technology such as the safety of pills and food dyes and airplanes and material strength and the reliability of brakes. But we have also social technology, such as the administration of certain educational techniques and tests, from the dissemination of textbooks to the administration of the celebrated, now thank-goodness defunct, British eleven-plus tests. We have town planning and zoning laws and building permits and so on, mostly antiquated. We have tests for sanity for compulsory admission to and subsequent discharge from mental hospitals which are meaningless and often barbaric. We have civil service exams and procedures for running offices. Of all the social technologies we have, most are unrepeatable. This is plainly shocking, and one need not be a sociologist to notice this. Buckminster Fuller was struck by the backwardness of building on land as opposed to the shipbuilding he had observed in the U.S. Navy; and this, he says in his *Autobiography*, is what has launched him on his now celebrated career as an innovative engineer.

The claim for scientific status, to repeat, is the claim for repeatability. We remember that technology is partly art but mainly science. There are claims for scientific status for eleven-plus exams which are supposed to sift the eleven-year-olds qualified for technical education from their peers qualified for higher or lower stations in life. There are claims for scientific status for tests designed to separate the mentally ill from the mentally sound. These claims are for repeatability. They are pseudo-scientific. They are, to use blunt, simple language, plain fake. They have to leave the scene without any trace.

Lies abound. Claims for liver pills, for cure of all psychosis by chemicals, for beauty hormone treatment for damaged hair and for aging skins, for anesthesia by acupuncture, etc. Some people get better when they take liver pills or sedatives or hormones; the claim for the scientifically attested usefulness of these is a lie since no repeatable observation reports are available to support it. By and large, it can be said, standards of repeatability are better observed in physical technology than in biotechnology, and better in biotechnology than in social technology. That is to say, claims for repeatability which are lies abound more in social or political or administrative practical affairs than in practical affairs to do with medication and with building and production engineering. When a new bridge collapses, it is more often due to poor administrative control than due to poor control of the physical conditions, though, of course, uncontrolled physical conditions – usually earthquakes – have their share as well.

This is not to say that engineers and builders and doctors are reliable. In fact, they are irresponsible too often, due to the professional conspiracy of silence, especially among doctors, and to gullibility and to the horrors of technical training, especially in medical schools, in internship and in residency. Yet, first, the credibility of claims of repeatable observation by the pharmaceutical industry and by medical research institutes is by far the highest, low as it surely is, and the educational and administrative and psychiatric fields have nothing to compare to them in level of credibility. Second, doctors are more reliable than bureaucrats; hospital administrations are less cruel and criminal and socially pernicious than teachers, headmasters and prison wardens and administrators. In short, these days we generally control things better than we control ourselves.

Are there social scientific facts at all? Most of the current discussions on the nature of the social sciences cover up just this question. Some say social facts are too complex; others say, sufficiently well-equipped, we can handle scientifically even the most complex facts. But on principle complexity has nothing at all to do with it: the conditions of repeatability (be they simple or complex) are stated in every statement of every allegedly repeatable observation – under such and such conditions, this and that happens. That many observations, both simple and complex, are both important and unrepeatable must be granted. But this is not of any concern for the discussion of the very possibility and of the very importance both of social science proper and of social technology proper. Are there repeatable social observations?

Economists say they have an easy and obvious example: all other things being equal, people prefer to buy more for less money rather than the other way around. This is not an observation. We all endorse it as sheer commonsense. But this is quite irrelevant. It may be true, it may be false, it may be too vague to be declared either; it may be just what the economist needs for his investigations and he is welcome to it. Yet "all things being equal" is not observable. Once we (a) specify which things we refer to and (b) deem them observable, we thereby render the statement in question an observation and it may, indeed, be claimed to be repeatable, and even true, and then properly tested and permitted to be used in technology. (This may be the case in our example, but only after the preference is deemed a market phenomenon, and statistical.) Are there other examples?

The tendency is, at this stage, to confuse repeatability with truth. There is a very celebrated classical observation report, all too often ridiculed, probably for very bad reasons. It says, "Decent women do not enjoy sex." Now one may wonder whether decency is observable at all, but the observation, made by nineteenth century British physicians, incidentally, meant by "decent" not a moral quality but a social status in nineteenth century Britain. So, yes, "Decent women do not enjoy sex" is an observation report, claims scientific status, and is not much different from "The sun rises in the east", except that it calls for the specification of nineteenth century British society made in observational language. This can be done: a fairly well-to-do, reasonably stable society will be good enough as a characterization of Victorian Britain – or any Victorian society, with "Victorian" becoming a generic name. Now "Relatively high-status women in Victorian society do not enjoy sex" is just like "The sun rises in the east". Both are repeatable, scientific, and, of course, unquestionably false. A rejected observation report wants a modification, a better substitute, we remember; the qualifier "appears to" will do the trick for both.

All this was only a means to remind us that a scientific observation need not be true. And so we can design many observations that are social, that are scientific, and that are obviously false. "All adults are married" is as good an example as any, though only when marital status is observable. But for technology we need corroborated observations – observations which we have failed to refute. For examples of such observations we usually go to psychology and fall on psychological observations which are not acceptable for the present discussion of social technology. Are there, then, any corroborated social scientific observations? Are there accepted standards for corroboration in social technology akin to those in physical technology? The answer to the second question is negative – and because the first question is commonly and erroneously negatively answered.

Do we have corroborated social generalizations? Yes. Example: All societies have the marriage institution and technologies and educational systems. There are many more laws, which, however, must be carefully worded though easily conveyed even in ordinary vague language. The death penalty is no deterrent, yet to make this a proper observation-report we must carefully specify both deterrence and the societies in which it is allegedly no deterrent since some people certainly do find it a deterrent for them personally, and in some societies the penalty – in the forms of feuds of all sorts – does deter. Similarly, jails are schools for criminals, but by jails we mean modern Western ones and we mean in a society like ours, which meaning must be specified in observational manner. Examples abound, but their wording is still wanting.

What has all this to do with the fine arts? All art is still the same as technique. As Ernst H. Gombrich says, the fine arts are merely outgrowths of the related crafts. He rejects the romantic theory of Art with a capital "A" given to men of genius alone, and he observes that the great masters of the golden age of the Renaissance came out of workshops. A medieval contract, he observed, calls upon an artist to produce a Madonna of the quality at least as high as a known Madonna was reported to have. In other words, the artist is, to begin with, an artisan, and in addition he is perhaps one with golden hands.

What we call the arts, or the fine arts – painting and sculpture and music – plus the applied arts – whether carpentry or advertisment – are all technology, and like all technology they can be prescientific. Otherwise, they are science and then they may grow a thin, but at times glorious, artistic veneer. Gombrich is not very consistent when viewing reproduction as not being art, then, even by his own light: all art is reproduction with variation anyway, as he himself says.

Thus, social technology is prescientific because it does not take care of the craft: its observations are not couched as repeatable observations, and consequently they are not seriously tested. Worse yet, when tested and refuted the refutations are plainly rejected, as the case with the still barbarian penal codes of most advanced Western countries amply illustrates. Once we allow our social technology to become a science, it may also develop and become art. The fear that social technology will be scientific and lose its artistry is rooted in an error. It may gain artistry only after it becomes scientific.

For the theory of art this has further interesting consequences. That art is both non-cumulative – non-repeatable – and cumulative – repeatable – Gombrich has noted; and he concluded that techniques are cumulative, scientific, not artistic, thereby depriving even the best artisans of their golden hands – quite unwittingly and quite inconsistently. Arts and crafts are in part unique, non-cumulative – art proper – and in part cumulative – science proper. Except that the reproducibility that makes art into science is, for craftsmen, less a matter of words and more a matter of skill. Yet this is a mere technicality. This also explains why, as Gombrich notices, though he admits inability to explain why, styles come and go. A style has its problems, and these are solvable first artistically and then scientifically. An artist who works in an old style is one reviving art where science is available: there is neither challenge nor sense in it. Also, this explains, perhaps contrary to Gombrich's theory (*Art and Illusion*), how come the individual who discovers a new technique (i.e., makes a certain task reproducible; i.e., makes a new piece of science) and the individual who finds good use for it (in practical matters and in aesthetic matters alike) need not be identical.

All this calls for readjustment of much that is said both in the philosophy of art and in the philosophy of technology, but most significantly in the philosophy of social technology. It is commonly assumed that in order to keep social technology as an art we must ignore social technology as a science. The contrary is true: only when, in our activity, we take for granted what is scientific in social technology, only when we have mastered the science of this craft, only then have we the challenge to develop it further and make it art.

5. CONCLUDING REMARKS

Intuition, inventiveness, spontaneity, and sensitivity, this is a complex of faculties to cherish. Modern society, like any society, has its own way of threatening this complex. The widespread educational attitude to art is elitist in order to preserve intuition and invention, and especially spontaneity; the tendency is to keep (some, highly selected) young people away from the humdrum of ordinary life, to let them stay away from the common herd, to prevent their delicate faculties from becoming gross through routine, to train them to shun techniques which render the hardly accessible readily accessible and repeatedly and excessively so. And this elitist sermon against routine is coupled with the severest routine artistic training, with endless five-finger excercises.

This attitude is self-defeating. Young people so trained are unusually crass and self-centered. The model for proper, sensible, sensitive faculties can be pure mathematics, which calls for large doses of intuition, inventiveness, and appreciation. The mathematician tries to routinize anything possible and delegate all that is routine to the robot. The fear that when all is routinized there will be nothing left for the imagination has been allayed by Kurt Gödel, but really we do not need him to know that if we can overcome the great dangers which mankind has set to itself right now, there is no reason to fear the exhaustion of our curiosity and wants by sheer mechanization. Art and technology need not fear each other.

CHAPTER 3

TECHNOLOGY AND SOCIETY

The Western democratic philosophy has been threatened in many ways and from many directions. Some criticisms launched against democracy are or were valued and worthy of acceptance, endorsement, and used as a means of improvement of democracy. Also, there are proposals of alternatives. Of the alternatives to democracy, some are not worthy of serious critical consideration, some are. All those proposals of alternatives to democracy are technnocratic: some technocracies are populist and egalitarian, such as Lenin's and Mao's, others are technocratic elitistic, whether elitist pro tem but egalitarian in principle like Soviet Russia since Stalin or elitist tout court like the individualistic extremists in the United States, known there by the misnomers "anarchists" or "radical right" or "libertarians". Does technology call for elitist government? Democracy, public participation in the process of governing or of running a state, political decision-making on an intelligent basis by the nation at large – has modern science-based sophisticated machinery made these outdated and antiquated?

1. IS DEMOCRACY REALLY NECESSARY?

Is democracy really necessary? Is it such a big thing? Should we defend it at all cost? What will make us give up the idea of democracy? Surely, if democratically we decide to turn into barbarians, then we'd better look for an alternative regime. We were once such a superior bunch that we felt democracy could never let barbarism take over. We are less self-assured since German democracy did give way to a government on the level of beasts of the worst kind – beasts armed with the most sophisticated technology available, technology which encompassed the finest precision instruments.

We have been warned. Over two thousand years ago we were warned. Democracy, said Plato, leads to total disorder and total disorder leads to tyranny. He was not just conjuring up phantoms to frighten us. He knew of the disorder and the tyranny that followed the Golden Age of Athenian Democracy. History gave us a repeat performance or two: a democratic revolution led to terror which led to tyranny in France as well as in Russia.

In Germany we saw the worst ever. Do we not play with fire when we insist on democracy and on further democratization?

An ancient philosopher, Thrasymachos, was quoted by Plato to say that there is no special merit in democracy: to begin with, the strong impose their will on the weak, and the weak wish to be strong and impose their will; at times they realize that they can only do so when they unite; when the weak unite and rule, their rule is democracy.

It is incredible how hard it is to think straight, how easily biases prevent plain thinking. Thrasymachos, Plato suggests, was an anti-democrat, and in error. Yet Plato was an anti-democrat no less than Thrasymachos, except that Thrasymachos was cynical about matters political, if Plato is to be believed. Plato recommended not the rule of the strong but the rule of the wise – of the wise Plato – for the good of the people. Cynical or not, was the point made by Thrasymachos true? Ben Franklin, one of the greatest, wisest, most humane of all democrats ever, said yes. He viewed democracy as a bunch of rascals: we must hang together or else we hang separately, he cheerfully quipped. How easy it is to love and admire his simple soul!

Things are not as simple. The argument against Thrasymachos is traditionally from the moral impulse of the democratic conviction, represented at its best in the democratic tendencies exhibited also in the ruling class and then symbolized by Mirabeau, a French nobleman of the mid-eighteenth century. Mirabeau and the moral sentiment are both irrelevant: cynicism may support democracy, as non-cynical Franklin argued, and moralism may oppose democracy, as immoral Plato argued. The question is, how good is the democratic regime, not how honest are its supporters.

Still, things are not as simple. The view of Thrasymachos is not only cynical, it also is a factual contention: democracy, he said, is the outcome of a wish to move from the bottom to the top. We do not know enough about the motive behind the democratic impulse. Historically, democracy was often supported by middle-class intellectuals and by better-off workers. Statistical methods concerning attitudes are young and their results do not agree with Thrasymachos. Also we do have many eminent cases of people moving upward because they could not tolerate oppression. In democracy, which encourages social mobility, these people tend to be democratic. But things are changing. From time immemorial, one major means of upward mobility was education, and when education was in the hands of the Church, the educated supported the Church and even used the Church for social climbing. Now that "tough" science and technology are the order of the day, educated people learn to talk "tough", use mathematical hieroglyphs to say simple things in obscure ways, extoll what they call excellence in education, and tend towards technocracy only to discover later, in retrospect, that technocracy and democracy are enemies.

Hence, if Thrasymachos is right, as he may very well be for all we know, then the same motive which traditionally created democracy will now create technocracy.

Lord Snow already defended the same idea without noticing it. It is impossible to doubt Snow's sincerity and democratic convictions. When he lectured on the two cultures, he branded artists as "soft", tradition oriented, and luddites, as opposed to scientists who are "hard", radicalist, and technologically minded, thus being inevitably optimistic or, as he put it, "future-oriented". Of democracy he said nothing in his lectures, so that it might go one way or another. He wanted to bridge the two cultures, but he was clearly on the radicalist "future-oriented" side. His main recommendation for such a bridge was the reform of the curriculum so as to make it science-oriented, so as to enable more individuals to joint the scientific-technological optimistic community. His model, of all things, was the Russian current educational system – as described by official Russian brochures which are radicalist and optimistic. It was later described differently in Solzhenitsyn's *First Circle*, which is traditionalist and pessimist. Even radicalists, however, today think Solzhenitsyn is more credible than Russian government public-relations officials. Yet the discussion started by Snow is still going strong.

It is time to notice how lightweight this discussion is. Science is no technology; technology has to be responsible, not be optimistic (optimism is not the same as responsible planning for the future); "hard" science and technology are not hard at all but muddled, confused, cowardly, reactionary – a veritable closed-society.

"Hard" science and technology is the ability of the uncultured individual with a smattering of some technical training to silence artistic interlocutors with mathematical symbols, to hurl them at the heads of hostile critics of science, to exhibit a blueprint to anyone who dares question one's importance, one's contribution to society, to progress, to all that is positive and good. "Hard" science used thus is nothing short of violence, although psychological rather than physical, and hence much less objectionable and perhaps also a bit less cowardly than physical violence – but not educated and not in the least intelligent.

Is Snow right about "hard" science being, as he observed, "future-oriented"? How many "hard" scientists are "future-oriented"? No one knows, but some of these organize from time to time in order to defend government-controlled research Stalin-style, attack democracy, and preach elitist technological education. Why? Why does radicalism attract preachers of technocracy?

George Orwell explained the popularity of Stalin among intellectuals in his day and age as the outcome of their power worship. And, of course, more intellectuals are prone to admire an intellectual bully than a political bully. Thus, at times artistically-humanistically minded intellectuals dread scientifically-technologically minded bullies, and vice versa. So the inferiority feelings of one group, especially the social climbers among them, support the bullies among them and may even be so bold as to become bullies themselves. And nobody notices that the bullies in both camps, both bullies with artistic-humanistic pretenses and bullies with scientific-technological pretenses, are cowards, marginal to their own groups, and guardians of closed societies – doormen and bouncers to restricted clubs. It really is all neither here nor there, since most of what goes on in the public eye is but froth. But this is not to say that the froth does not count. It counts because it diverts interest and concern – away from the real problems of the day. For, the bullies manage to entrench accepted standards, even when these are poor. The pride that a scientific and technological expert may have in this inarticulateness, as described, for example, in Aldous Huxley's *Suddenly Last Summer The Swan Died*, is not necessarily widespread, and it may be a minor and understandable matter that the articulate scientist and technologist is less willing to exhibit his strength than the inarticulate one is willing to exhibit his weakness as if it were strength. As a result, education in science and in technology does not include the best training in articulation. Having no courage to complain about their education and to try to alter it, victims of this educational deficiency find it easier to hide behind the bully who brags about it, though they know he is the proverbial fox who, having lost his tail in a snare, pathetically proposes taillessness as the newest fashion straight from Oak Ridge.

The situation is the same in the arts as in the sciences: the artistic bully resembles the scientific bully; yet it is the scientific bully who is our concern here, since he threatens democracy by proposing to replace it with technocracy. We are thus forced to choose between developing means of democratically controlling the bully or submitting to his technocratic bullying. All this, however, is no argument in defense of democracy, only a democratic response to technocratic bullying. For, democracy was criticized by Plato as the regime that allows the rise of some bully to power, and this criticism has been proven in experience. It is now agreed that democratic public opinion can lose its ability to control governments. People are all too often tired of democracy when the public is hoodwinked and fooled and frustrated. They refuse to see that democracy is the education of the public to tend to its interest. They may still recommend that a wise ruler should be appointed to run public affairs until the public learns to practice democratic control. This, more or less, was a modern variant of Platonism – the idea of guided democracy. It often enough does not work since the wise ruler tends to declare the public still not able to learn to control the government as long as the government is his.

It is because democracy is educational and egalitarian and participatory that technocracy is advocated by uneducated people who have narrow skills and who defend their narrowness as "hard" science in an effort to appear superior to others, less fortunate than them in their skills, particularly those who are also more fortunate than them in having developed tastes for culture. The result that the "hard" and the "soft" have no part in each other's activities, and so together they advocate the confinement of each to his special proficiency, pretending to agree that it is best to leave government to the expert governors. Such a system, were it possible, would disintegrate for want of an overall plan for survival unless a harsh ruler takes over. This argument is fatal. Of course, it is not new and not without a response. The response is that we can have long-range, large-scale planning experts to supply the expert governors with long-range, large-scale plans intended to keep society intact. This makes the super-experts the real rulers, however.

Before giving government over to the super-expert long-range, large-scale planners, we may wish to see how well these can plan. We know how poorly governors govern without the aid of planners. Let us look at the planners, then, the last hope to intelligently justify technocracy, the only serious competitor to democracy.

2. LONG-RANGE, LARGE-SCALE TECHNOLOGICAL PROJECTS

The idea of a technocratic society can be critically examined with relative ease due to the fact that technocracy is already evolving and we can see how it works. The rise of technocracy is a gradual process, due to the rise of modern technology and the political power it wields, and due to the absence of any social theory of, or even research into, the dangers of technocracy or the need of developing a social technology adequate to the developing physical and biological technology. (Marx was right in seeing the growth of social and political thought as an after-thought to technological growth; he was in error in saying that this is inevitable and that successful planning for the future is impossible.) The technocrats are those who created the social and political bodies needed to serve the implementation of new technology. They usually were unscientific organizers with a smattering of knowledge of the scientific technology whose implementation they served. But they soon had to hire experts; the new expertise is known since the turn of the century as scientific management. This heralded the evolution of a scientific social technology – as opposed to a physical one – with schools of business administration soon occupying in the university the same position as schools of engineering, and they emulated them as best they could. Scientific management was, from its very beginning, pseudo-scientific social technology.

Things got out of hand when business administration tended to encompass all administration. When scientific technology became so very important and all-embracing, the tendency was indeed this. The inability to define technology, the inability to confine technology to the mere provision of physical and biological tools for the control of our environment, and the lack of any adequate consideration of this inability forcefully invited the growth of a class of technical experts who administer and control increasingly large portions or aspects of our lives. Unless we effect a radical alteration of our system, especially of our educational system, we are going to see democracy turn into technocracy, with elected officials losing their hold and resigning the helm to the appointed officials who should serve them. The officials' way to the top will be paved with technological achievements or with the pretense of them. We already have increasingly large portions of our public life designed, planned, and executed by experts whose authority rests on impressive technical achievements or on claims for them. *Parkinson's Law* is hard at work here: committees spend but a few minutes debating the "hard" aspects of a plan to erect a multi-million dollar nuclear plant designed by experts, simply because they would not be able to discuss the plan intelligently. Parkinson also observed that such committees spend hours debating – still the "hard" aspects of – plans for building a cheap bicycle shed, since they comprehend them. But this aspect of committee work can easily be improved upon. The question is, how do the experts decide – since committees only yield to them – on plans to build nuclear plants?

The answer rests on what is known as technological forecasting. It is important to observe that technological forecasting is an essential part of proper technological planning and assessment. Technological forecasts are thus imperative. A modern government which does not try to assess demographic trends, for example, cannot begin to plan responsibly. And, of course, such assessments can easily be found often unreliable. How will technologists or technocrats meet the challenge of responsible planning?

The problem at hand is that of the reliability of long-range, large-scale planning. That is to say, what are criteria of reliability, and what to do with unreliable criteria (and plans) in the absence of reliable ones. The reliability of long-range, large-scale planning – macro planning, for short – and of macro forecast are intertwined: when we can effectively forecast a calamity we plan counter-measures and effectively forecast their effectiveness, for example. That macro forecasts are too often poor is proverbial. The National Planning Association in the United States of America issued in October 1959 a Staff Report, Planning Pamphlet No. 107, Washington, D.C., by the tell-tale title, *Long-Range Projections for Economic Growth: The American Economy in 1970*. A few years later, in 1966, E. Kirby Warren, in his *Long Range Planning, The Executive Viewpoint*, praises this pamphlet in the light of experience. Yet, even though half way to the target time the projections still looked not too bad, the forecast can now be assessed with sufficient precision, and its usefulness can be studied. Was it used, and did those who used it benefit or lose from their reliance on it? Even if we knew, we would not know whether the success was due to wit or good luck; or whether the failure was due to folly or to ill luck. In such circumstances, we can say, forecasting itself is pointless: it is a sheer game of chance, mere airing of possibilities, mere flights of fancy.

One might feel that the skepticism pronounced here is of the philosophic style, of the kind that philosophers professionally and habitually engage in, when they are willing to doubt even that they have arms or legs. Clearly, however the philosopher's doubt is judged, in the present context we can take the reality of our own arms and legs to be certain. But the doubt about planning is of the commonsense variety. Thus, Aaron Wildavsky asks, "Does Planning Work"? in *The Public Interest* (24, 1971, 95–104) (hardly a philosophical journal), and answers: "The record of planning has hardly been brilliant. For all know, the few apparent successes (if there are any) constitute no more than random occurrences. . . ." Also, the same passage is quoted by E. S. Quade in his *Analysis of Public Decision*, N.Y., 1975, p. 251, without disagreement.

The radical change in macroforecasting occurred in 1973, with the oil-crisis so-called and the energy-saving programs. The major factor is, of course, the fact that the crisis invalidated all extant macroforecastings as all-too-inaccurate. Second, the crisis was artificially created when an irresponsible, helpless United States administration permitted oil companies to run wild in the hope that the damage this would cause Japan would be greater than it would cause the United States, so that Japanese textiles and other manufactured goods would cease flooding the United States market. In the ensuing chaos many macroforecasts were circulated, some officially, some leaked to the press as if they were confidential – as authoritative scientific forecasts, as government ones, or as ones sponsored by the industry. Since then it became apparent that one must issue macroforecasts regularly and correct them regularly. This should become standard practice. The latest corrections of macroforecasts of energy consumption – especially by the European Economic Community – have been altered radically in very short periods. It is now becoming obvious that macroforecasts are both extremely unreliable yet extremely important all the same.

Nevertheless, we may wonder why there is such an utterly negative attitude to all past macro plans. Let us glance, for a while, at part successful programs. The advice to start at existing [successful] programs, has been made much of, by the way, since Herman Kahn, the master macroforecaster, advised planners always to pretend to be working on existing plans and call their plans and programs new only when these are glamorous (*On Thermonuclear War*, Princeton University Press, 1960). Not that it makes a difference: we do not have criteria of novelty here. At times we can nevertheless take clear-cut new macro plans, such as new town plans. These, unfortunately, are very hard to assess even as to their initial purposes, with exceptions, of course, such as Haussmann's Paris and the British new towns. Even this is questionable, since modernizing Paris had not only the prime military purpose of preventing barricades and the economic purpose of attracting capital, but also the immediate purpose of supplying large-scale employment and the overall political purpose of changing the quality of life in the capital. For really simple innovative macroplans we must take examples such as the Great Wall of China and Hadrian's Wall, the Suez and the Panama Canals, or the Don-Dnieper Canal and the Donau-Main Canal. It is obvious that such macroplans are embedded in larger macroforecasts, of the future of empires, of the future of industries and trade routes, of the future of transportation methods. Yet in almost all such cases, the macroforecasts are hardly problematic, and because they are embedded in every aspect of life.

It is well worth stressing this. We all assume that the sun will rise tomorrow. Philosophers and scientists tell us how this assumption is scientifically justified. Laplace issued two proofs for the stability of this forecast, the Laplace rule of projection from long past processes to their brief continuation, and the Laplace proof of the stability of the solar system as a whole. Both of these proofs are problematic, to say the least, and undoubtedly not easy to follow, so that at least most readers have to take them on trust and not on the basis of proof. Often enough textbooks refer to them without discussion, offering only the end result, not even the outlines of the proofs. In Laplace's time Kant proved the instability of the solar system. The proof is easy: energy dissipates from it – e.g., due to tides – and hence the system cannot go on unchanged. This proof became obvious when the source of solar energy was better understood: we now know that the fuel of the sun is nuclear and hence the sun, being a nuclear pile, will one day explode. It can happen tomorrow, we are told. The probability that it will is small, we are comforted; but what will we do if there was a high probability that it will explode tomorrow? Just what we are doing today. In other words, some suppositions are taken for granted out of no choice: without them planning makes no sense, micro and more so macro.

This looks like a transcendental proof: we need these assumptions; hence they are true. But not so. For all we know they are false. But decision theory tells us that if the morrow does not come, the gain or the loss is small, whether we plan today or not, whereas if the morrow does come, the loss due to no planning is big and the gain due to planning may be large. Hence we better plan, come what may. Without plans, says Neville Shute (*On The Beach*), life has no meaning.

This is a characteristic of planning which can be generalized: we can find suppositions which are necessary for our plans and ones which, if false, make null and void the very attempt at planning. These are, then, blanket suppositions par excellence.

The reason that tomorrow's sunrise is such an excellent example of a blanket supposition is that our very survival – both as a race and as individuals – is tightly linked to it, and survival is the initial purpose of all plans. We can go further and say that any initial purpose gives rise to some blanket supposition. Moreover, once a large-scale plan to satisfy an initial purpose is taken for granted, assumptions as to its success to a minimal extent are blanket suppositions to any small-scale plan embedded in the large-scale plan – the secondary purpose is dependent on the primary purpose, and, in the circumstances, it depends also on the specific plan to satisfy the primary purpose. Hence, the way to make a plan viable is to discover its blanket suppositions, including the anchoring of it in a more general plan. In common parlance, tactics receives its significance from strategy, never the other way around. Practically never, that is.

It is therefore important to observe that some blanket suppositions, say, that the sun will rise tomorrow, apply to all of us. We can have blanket suppositions for nations too, and even for individuals. Dictators are opposed to birth control, for example, because they plan to have cannon fodder twenty years later: it takes about that amount of time to produce a full-grown soldier. This is based on the dictator's deep convictions, especially his deep conviction regarding the need his country has for a dictator like his own majestic self, able to lead it to war even twenty years later.

All this shows very clearly that the most important aspect of our initial purpose in the study of long-term planning and forecasting is to examine blanket suppositions. The public awareness of this situation is what has given the ecological movement its impetus: the movement got going only after the blanket suppositions it was attacking for decades were empirically refuted in a most glaring manner: the globe is now polluted sufficiently to make it clear to vast publics that going on as we did till now will soon exhaust all our resources, including, particularly, fresh water and fresh air. What makes macroforecasts and macroplans so very urgent is, indeed, the feeling that the global forecasts for doom are not so unreliable as we would wish. Thermo-nuclear war, population explosion, and the irreversible pollution of our ecosystem, are serious risks; their combination makes disaster almost inevitable, unless we do something. But what? How?

By now the initial questions presented here have been left far behind. The questions were, what are the criteria of the reliability of plans? and, what do we do when our plans are very unreliable? The present perusal of these questions has ended up with the obvious observation that we have no global plans, whereas the safest plans are those embedded in bigger ones. Yet this defect is very common in the literature. Why is the defect ignored?

Let us consider the view that the defect is not important, since for blanket suppositions almost any plan will do, and we always have plans whose level of reliability is exceedingly low, or as low as possible. We can consider the following plan as an example – the plan to save the world by calling everyone to join in the program for worldwide prayer and fasting. This plan is just too unreliable and will offer us no interesting or useful blanket supposition.

Hence, we require overall plans that have prima facie reliability; we can then examine their reliability by better means than mere initial impressions. But plans of this sort are simply not in our possession. This, then, calls for an analysis of the situation. We can examine, for example, the failure of ecologists to arouse interest in the welfare of humanity as a whole before the last decade or two. We will find that they have no special blanket assumptions – unless they maintain blanket assumptions rejected by most people. Can one work with no special blanket assumptions?

Attempts to work with no special blanket assumptions abound. Their results are appallingly poor. Analysis will easily show that this poverty is not accidental. The first thing one may notice is that such works almost exclusively concern microplanning, even when considering large-scale decisions, such as Detroit's decision to stick to constructing large automobiles. Why? What went wrong with their plans?

Planning is usually micro in the sense that planners take too many blanket suppositions for granted on the strength of inertia even when intertia regularly fails. The fact that so much is taken for granted uncritically and naively and unthinkingly is covered up; the cover-up is an enormous pile of information including the finesse in the study of details, supported by diverse technical means, such as consumer and market research, and the analysis of data with the aid of high-powered mathematical tools, including computers, systems analysis, statistics of the most arcane sorts, and more. What this insures is that one Detroit competitor will not likely have an edge on another but that Detroit as a whole may go under. And, indeed, when Chrysler collapsed it was a sure bet that its competitors would follow suit, all lies to the contrary notwithstanding. But this is the wrong story. For, better techniques were vailable to Detroit had the American planners not been so insular. And then, we would be able to say, especially in the light of the automobile industry becoming increasingly multinational, no competitor will be likely to have an edge, home or foreign, but almost the whole industry of the Western world may collapse together.

The question, what to do when there is no alternative plan, is often answered by the scenario theory, so-called: describe what you think is likely to happen without the existing plan. The moment we have a technique, we can devise alternatives to it. Indeed, standard alternatives to the scenario are operational gaming, which is scenario writing with the use of illustrations rather than verbal write-ups, perhaps with special emphasis on some mathematical techniques and illustrations; brainstorming, which is but any old attempt of a group to stimulate itself to speak spontaneously and almost thoughtlessly in the hope of hitting upon any new idea; and the Delphi technique which, to the contrary, is the eliciting of clear-cut responses from many experts in writing, taken as anonymous and statistically analyzed.

The amazing fact is that these few primitive techniques occupy a full-blown, serious literature. The poverty of ideas is bottomless. The Delphi technique, which uses a questionnaire as a framework, is particularly futile. Independently of questionnaires, it is hard to deny that the experts consulted share a framework. If one of the consulted experts transcends the framework, he is more likely to be misunderstood and squeezed into the framework, or simply ignored, rather then noticed. Statistical methods ensure this: after all, statistics is meant to iron out differences. And why should he be noticed? Most who break out of the framework are, indeed, quirky and often silly. Feedback, i.e., recycling of results, adds a bit, but is too mechanical to help much. The most incredible fact about the Delphi method is that having tested it we come up with an incredibly poor result: some information – factual as well as theoretical – is quite reliable; and when expertise is good, experts share it. Hence, only when using reliable experts in order to weed out what is not reliably accepted, the Delphi method is handy. But the trouble is that we do not know what class of experts or what field of expertise is reliable. Only after successfully using the Delphi method in order to establish expertise can we use experts by the same method in order to exclude unreliable items. But, once the method has established expertise, it becomes hardly necessary: consulting two or three experts face to face will do equally well once it is established that they are good. But, and this is the tragedy, by using the Delphi method we get some blanket suppositions entrenched whenever they are shared by all experts – or whenever they are shared by all experts in Detroit and the questionnaires are administered to local oracles only. In other words, the expertise that the Delphi method established is the fact, if and when it be a fact, that all acknowledged experts share some blanket supposition, even if it is that only prayer will save the world. What, then, is to be done?

E. S. Quade's 1975 book endorses the negative view of all past planning techniques. Quade himself hopes for improvements. He words his advice in very clear suggestions (275-277): "Give policy analysis a recognized place in policy making", "provide training", and "support and maintain policy analysis capabilities". No doubt, this is excellent advice as far as it goes, which is, alas, not very far at all: before we provide training, we need good plans for training and good likelihoods for their future success.

The scenario method calls for no comment; it is neither new nor very useful. The same can be said of the method of brainstorming, except that it is exciting and so all to the good. It runs contrary to the traditional methodological taboo, shared by so many superstitious experts on scientific method, against wild imagination. This is why the sessions are organized playfully, on the famous principle: smile when you call me a liar!

The view of scientific method current in the scientific world, and more so in the technological world, is very authoritative, yet excessively naive: science begins with an enormous fund of information, and slowly evolves its theories from the facts and further bases them on further facts. Thus there is popular aversion to the study of blanket assumptions. In underdeveloped fields the blanket assumptions are smuggled in unnoticeably and unwittingly and uncritically; in developed fields they are introduced by some intelligent bold thinkers. The field in question here is underdeveloped. Further, the view of the social base of science current in the Western scientific world, especially in the Western technological world, is excessively simplistic. It is the view that the West makes technology develop as rapidly as possible; Russia is the runner-up. That this is excessively naive is obvious: the ecological crisis, the population explosion, and nuclear proliferation are proofs that we do not have a clear idea of rapid technological progress.

Nevertheless, there is a problem lurking here, which might be of some interest. The success of modern technology began when physical technology simply integrated into the pre-existing social fabric. Then, in the industrial revolution, technology simply upset the system so much that Charles Dickens could describe, and Karl Marx could reasonably analyze, the strain on the system. Marx predicted that the very growth of technology will force a collapse of the system; as a result he expected society to revolutionize. The immanent collapse, however, created initiative to reform the social fabric. Short-range planning techniques were invented. Their extension to long-range problems is the fiasco for which the human race may yet pay by the highest price of extinction.

3. SHORT-RANGE CYBERNETICS

The purpose of the present discussion is to describe and explain the new branches of technology which have to do not with material achievements of technology but with the man-machine systems, or with systems as such, since the achievements of this kind are what give technologists reason for their development of a taste for governing and controlling people, an excuse for this taste from allegedly "hard" science, and the hope that government can be run by "hard" techniques which render democracy unnecessary and impossible, since they are both reliable and not accessible to ordinary people.

Incidentally cybernetics, systems theory, games theory, decision theory, and so on, are both accessible to ordinary people and can better serve as introductions to mathematics and mathematical thinking than Euclidean geometry and traditional high-school algebra customary in Western high-schools today.

It is better not to study the history of cybernetics, since its prehistory is complex and still subject to exciting studies. Its birth date is World War II, when cybernetics, games theory, decision theory and operational research were all developed.

Operational research was created in the British forces by people who noticed the need for overall planning and for coordination between sectors of fronts and of military industry and of military manpower, etc. The operational research students made it a profession after World War II simply for want of jobs. Later their results were incorporated in systems analysis studies and thus in cybernetics. Cybernetics became, as intended from the start by Norbert Wiener, who coined the term – cybernetics is the Greek version of the more familiar Latin governing – and whose 1948 book *Cybernetics* defined the field broadly as that of information and control and added at once games theory and everything else he could add. Anyway, the terms "operational research" and "systems analysis" will be avoided here as much as possible, and the other terms will be explained later on.

The main characteristics of cybernetics is that it is seemingly "hard". It seems to be hard for two good reasons – the bad reasons are best ignored. First, that its activities make use of high-powered mathematical knowledge. For the understanding of an intellectual activity which involves high-powered mathematics there is no need to be in command of the mathematics involved, much less to be able to calculate: once one realizes the ideas involved one can request the aid of an expert mathematician – usually a professional, who can do so for a reasonable fee. This rule is quite general and classical. Even the simplest sort of problem in classical mechanics, which all physicists should know as a matter of course, involves mathematic techniques beyond the competence but not beyond the comprehension of most physicists, past and present. The second reason that makes cybernetics look "hard" is that the simplest sort of problems may easily become difficult to handle once they become complex and they become complex very quickly upon reiteration. Take a simple road map, and consider how one works out networks and rules for the choice of a road from one place to another given some of the usual constraints drivers usually take into account. Just writing out all that this may involve gets quickly out of hand. Drivers consider such problems piecemeal, and if they are in trouble they consult advisers at the automobile association. The advisers, too, consider the problems piecemeal, though some of the solutions are written down for them by bigger experts, who may also have taken them piecemeal. But these super-experts may approach the problems systematically. There is no more to the systematic approach than what the ordinary driver knows, yet it can be quite awesome. Many problems of traffic flow may enter the picture, especially if the constraints imposed on the solution force the driver to move through thick traffic during rush hour. But for the invention of the computer, many of the solved traffic flow problems, simple as they are in principle, would not have been solved, so complex are they in practical detail. Indeed, when these problems were first tackled, computers were still scarce, and so most of them were allowed to be left fallow, and some of them were later picked up when computer power became more accessible.

One can complicate matters more by introducing to the considerations of the road map problems also some probability considerations. But we need not enter into all this, since it is easy to find material on such matters elsewhere. Suffice it to say that what has been said thus far, though trite in principle and so it should be common knowledge, is in detail both complex and involving exciting high-powered mathematics and exciting mathematical problems – including the limits to solubility-in-principle, and the limits to solubility-in-practice.

This means that one who faces such problems – as a driver, an adminis-trator, or a ruler – may present the problem intelligently to a mathematician who will help solve it for a small fee. He may, however, present his helper with an additional task: he may present the raw material to the mathemati-cian and ask him for instructions. The liberal mathematician should then try to find what his client wants; alternatively, the mathematician can try to decide what is good for the client. In any case, this added task will enable the mathematician to call himself an "Operations Researcher" or a "systems analyst" or a "scientific administrative consultant". He will be on his way to becoming a technocrat and then run your affairs even without being approached in the first place.

Back to simple principles. One of the most astonishing mathematical innovations of this century is the flow chart. A flow chart looks like a schematic road map, the like of which London Transport has, which has been emulated by many a town's public transportation authority. A flow chart can say, go from A to B, from B to C and so on. Or it can say, at B you have a choice between CA and CB, and road CA should be taken when X is the case and CB when Y is. Also, a flow chart can say, into factory A you introduce raw materials a, b, and c in a given ratio as well as energy and manpower at a given rate, and out of the factory you get finished products d and e at a given ratio. One can then assess the cost of what goes in – the input or the imput – and the income gained by what goes out – the output – so as to decide the advisability of operating the factory. Also, of course, the factory can produce different products, or different ratios of products, also the output can be an unfinished product or an output that has to become an imput somewhere else. All this is so plainly obvious, one wonders what is the good of saying it all?

It is amazing how much energy is wasted by the disregard to this question. The answer to this question is obvious once it is known, but otherwise one can really be puzzled and disturbed by it. What are flow charts good for?

Flow charts are good for two things. First, complex flows are best understood with their help. They are not essential: one can study every step of a flow chart and develop a general idea of it without drawing it; yet drawing it enormously increases the ability to develop an overview. There is one other and particularly useful function of flow charts – they make it easy to program and to control production processes. In particular, they are useful for the design of the very revolutionary means of control – the feedback mechanism, the heart of cybernetics.

Feedback mechanisms are very, very simple in principle and may be highly sophisticated in some modern servo-mechanisms (= robots). The definition of a feedback mechanism cannot be generally given since the same feedback process can be effected by diverse mechanisms of diverse sorts. But the feedback process itself, the system of control by feedback, is very easily defined: when a part of the output (= product) is used as imput (= raw material) for the purpose of control, then we have a feedback process.

The best-known feedback mechanism is the thermostat. A thermostat consumes part of the heat which a heater produces, so that viewing the thermostat as a part of the heater we have some of the product – heat – fed back into the heater. When too much heat is produced, the thermostat cuts or reduces the flow of fuel consumed by the heater; once the temperature falls the thermostat raises the heat. Many kinds of thermostats exist, each utilizing mechanisms affected by heat in known ways, which ways are then utilized as controls. The word "feedback" is often used as a synonym for control. Thus, the flow of information the company uses to control production is often called feedback, even when the information is economic and concerning foreign markets. When workers eat products to check quality, it is, literally, feedback; when consumers eat samples to help a consumer study of a market-research company to gather some information, this is hardly feedback. Even when a worker eats a piece of a product it is not quite feedback – unless the energy from his food counts as imput! When a quality control unit of a plant uses a product of that plant, this certainly is feedback, though hardly automatic. When the quality control worker is replaced by a machine, the feedback control is automated, so that we can now speak of a feedback mechanism proper.

A thermostat raises low temperatures and lowers high temperatures, thus keeping the temperatures static. Anything which keeps something static by feedback control is called homeostat. Homeostats are the simple negative feedback mechanisms, since they correct fluctuations by reversing them, thus keeping systems stable. Positive feedback mechanisms increase fluctuations, thus causing some explosive or implosive tendencies. Complexes of feedback mechanisms are called servo-mechanisms.

A complex automated or semi-automated factory can be described by the help of a flow chart, which can chart alternative production programs, controls, feedbacks, imput-output sub-systems, flows of imputs and outputs, queuing and storage techniques and more, and where each detail can be described by simplified equations and the solutions help program the whole system – including constraints on the system which can relate to profits, risks, etc. This way one incorporates many branches of new technology – industrial engineering; queuing and storage theory, (linear) programming, and, quite significantly, decision theory.

Decision theory presents a list of alternative hypotheses and pairs of lists of favorable and of unfavorable outcomes of acting on a hypothesis – once given that it is true, and once given that it is false. In some cases, acting on a hypothesis is profitable if it is true and hardly punishable if it is false; in other cases the profit is large but not likely, whereas the penalty small but very likely. The cases of the first kind are hardly problematic, the cases of the second kind call for policies: do we care more for maximizing profits or for minimizing losses?

Most known problems of decision theory are as yet unsolved and perhaps unsolvable. Yet the very earliest developments in the field were so powerful that the whole field of decision theory was declared a military secret by the military authorities of the United States until World War II was over. In particular, the method of control is linked to decision theory most powerfully. Whenever we have a quality control problem, we can use decision theory to decide how to act given that quality be this or that, that the cost of quality control and its benefit are different in different circumstances, and that we judge the circumstances correctly. More generally, control is, in one way or another, the purchase of information. Whenever we have a tough decision problem, we can ask, is it worth making a crucial experiment between the present competing hypotheses? In war time this means, Is it worth sending a patrol before the operation?

What should be emphasized is that all that has been said here thus far is short-term and embedded in specific contexts. The context, the given framework, the axioms of the system, are not questioned and there is no room here to question them. This is particularly clear in decision theory, where competing hypotheses and estimated probable cost and benefit of endorsing each of them when true and when false are all highly context dependent. So is the economy and the technology given in the case of cybernetics. There is no context-free cybernetics, and no context-free decision theory. But since both these theories present the context in an opaque manner (i.e., without specifying it at all), the situation baffles the novice. The novice will be easily unbaffled were this said explicitly, but it hardly ever is.

Why? Why is the implicit context dependence not stated explicitly? Perhaps because textbooks are always magisterial and never explain. But perhaps also because technocrats like to pretend that they solve more problems than they do. Technocracy is impossible when macro plans are beyond the reach of the technologist, and so he pretends to have macro forecasts and plans, and he pretends he solves context-dependent problems while allegedly controlling the context independently.

In some cases this can be done. Rather than calculate the speed of gravitational flow of water from a reservoir – an exercise we can do with not much difficulty – we can simply make the outlet from the reservoir so big as to keep to a reasonable minimum the time required for it to drain empty. Rather than ask what heavy cargo we may allow to cross a bridge, we can relatively cheaply reinforce it far beyond the strength ever required. The cost of building a bridge is such that it makes much less difference to it whether it is reinforced to excess or not, than whether it will survive an earthquake or not, however improbable the earthquake might be. A moment's thought will reveal that this cannot be done in all circumstances – what is a stake in one circumstance is a dagger in another. Sir Walter Scott's *Talisman* begins with the description of a medieval British knight's armour becoming a death trap in a combat under the blazing sun of the Near East.

This discussion is too long and too burdened with information already, yet it may be necessary to add to it one more detail – from information theory this time – not because of its philosophical interest, and not because it is part and parcel of modern sophisticated technology, but because it is often presented as context free. In a sense it certainly is context free, yet the reliance of it out of context is dangerous and even silly. It is, indeed, one major presupposition of information theory that in principle any item of information can be reproduced to any desired accuracy so that it is in principle possible to reproduce any item of information, including, say, any picture of any old master. This, presumably, is trivially true in some sense, if we really ignore the question of cost and permit as an expense even research into problems unsolved as yet whose solution is required if the reproduction is to succeed. It is hard to see how information theory as such, which deals with the communication of signals proper, has to do with context-free or with long-range planning, yet the cleverness is impressive with which information theory permits us to discuss communication while ignoring the question, in which language it is, or the question, in which channel of communication the message is transmitted. The fact that this can be done shows the power of generalization and abstraction in diverse unexpected directions and this power may be impressive enough to merit technocratic promise of orderly control of anything. Indeed, in Fred Hoyle's scientific utopia, *The Black Cloud*, a story is told of a catastrophe which is forecast by a bunch of scientists who devise an international information-processing center and with its aid gain enormous power of control over the whole globe.

Information theory is not context free but refers to opaque contexts, and these must be explicitly supplemented for the theory to successfully apply. It is well and good to speak of words in any language and of their frequency, and to argue, as information theory does, that the infrequent words are better repeated to reduce miscommunication. (The whole message repeated is a better but costlier means of avoiding miscommunication in the physical sense of the word: the more manuscript copies of a book we have, the easier it is to reconstruct the exact original. But in another sense a rare word repeated is easily accessible whereas different replicas are less easily available or even handy. Hence decision theory enters the picture again at this junction.) But in order to be at all able to apply this knowledge, one needs a concrete language and concrete information about the relative frequency of given words in that language.

This makes information theory without statistics inapplicable, and statistical hypothese are made within alterable contexts. Hence, the more specific the context, the more useful information theory is. The specific context and the extensive use of information theory together bring about two inseparable results, one excellent and one terrible. The excellent result is the smooth operation of the flow of information: noise, miscommunication, disturbances, are reduced to a minimal level; miracles like the reconstruction of a wonderful but crumbling Renaissance fresco can best be effected with the judicious use of information theory. The use of the word "judicious" is crucial here. For together with the wonderful result comes an awful one: any unusual and exciting message can easily be slightly distorted and thereby made conventional and common place; and the way to reduce noise is exactly the same as the way to distort an exciting message: statistically the unusual is the unusual, be it noise or innovation. Hence, the over-application of information theory to the reconstruction of a Renaissance fresco will make it typical to the class of messages it is deemed to belong to, and its unique qualities will be smoothed out: its magic will be forever lost.

There is a profound moral to this.

4. THE UGLY FACE OF TECHNOLOGICAL BUREAUCRACY

The attraction of technocracy is not necessarily the hope of the technician, the engineer, the scientist, or the science teacher, to be more successful in a technocracy than in a democracy. Rather, he knows that his work organization is not democratically run, whether it is a workshop, a corporation, a research institute, or a university. And he can plainly see that these organizations which are scientifically-technologically advanced are much better organized than the country at large, its political apparatus, or its civic or commercial bodies. (In Israel the best organized body is the military-controlled, state-owned aircraft industry, not the parliament.) In the 1930s many an honest and decent intellectual went Stalinist in the hope that Stalin's cruelty would prove to be no more than a regrettably needed short-cut to technocracy.

There are two pictures of technocracy, utterly unrelated, naive and realistic. The naive picture is of a scientifically and efficiently run society. It is not empirically criticizable unless taken as a detailed plan. As an aspiration it is criticizable on moral and aesthetic grounds, not on empirical ones. The realistic picture includes ideas about resolutions of conflicts of interest. This is possibly criticizable – if put forward in good faith. It is not criticizable at all without good faith since it is otherwise misleading and should not be debated. Here are the two pictures of technocracy, then.

The naive presentation of technocracy begins by exposing the ills of democracy and analyzing them properly as rooted in the ignorance of the population. It continues with the resultant possibility for knaves and fools who are ambitious enough to get to the top, in democracy and in other regimes no less, through lies, intrigues and pettiness.

So far so good. Next comes the idealized picture of the scientist, which is not a lie but merely an error and an exaggeration. It describes the scientist as a combination of a technologist and a scientist, and as a high-minded and critically minded expert who knows the abilities and worth of his peers.

So far not so bad. Now all we need in addition is the readiness to introduce to all walks of life the high standards practiced in the commonwealth of learning. This, briefly, will expand the commonwealth of learning to cover all social roles and render all social activity scientific technology. The key expression here is not necessarily the abolition of democracy (since one may admit that no regime has been able to do as well as democracy). Rather, the key expression here is the rendering of all activity scientific – which means that society will become the conglomeration of scientific industry, scientific management, and university – with education and research blooming everywhere. Political controversy will thus be scientifically settled.

This image is, to report an empirical observation, immensely popular among educated top members of technological, organizational, and educational bodies, and it is acknowledged as intellectual elitism and defended on the ground that we cannot let students run a university, workers to run the work place, and so on. Elitism, these people say, is merely the equality of opportunity.

Technocracy, like democracy, begins at home, and if we wish to overcome technocratic tendencies, we must democratize the university and the work organization and the public service. (Harold Lasswell attempted to democratize even mental homes. There were attempts to democratize jails. But since both need vast reforms, there is no need to talk of them in this context.)

The moral and aesthetic criticism of elitism is hardly in need of articulation. Some people like their doctors not to transmit to them any information and merely to prescribe to them, to operate on them, etc. This leaves to the expert all responsibility for one's health and welfare, including the decisions as to how to maintain it and at what cost. Responsible patients, of course, deem the doctor an expert adviser and an agent, not a decision maker. Suppose every aspect of one's life to be run by others with the possible exception of one's own specialty. In that case one's judgment within one's specialty will reflect current professional views. No more need be said: let the imagination take control and everyone can have a private picture of technologically controlled utterly heteronomous way of life.

The imagination will, of course, sooner or later break down, since all aspects of one's life are governed according to this picture by technical considerations, leaving no room for politics – democratic or otherwise. Parts of one's life are governed by all sorts of public administrators who might decide for one whether one should get married or not, for example, or whether to change one's nationality. Politics will not enter the picture.

What is politics and who is an expert politician? The moment we find a political expertise and rationalize it, it becomes an administrative profession and we make it scientific. Politics as the art of governing a social or a political body is politics no longer: it is public administration. Politics may be the art of decision making. We certainly do not consider all decisions political, and with no political institutions proper no decision is political: the very difference is between administrative and political decision making, and it pertains to public administrations and to political organizations. The naive picture of technocracy does not recognize any political organization and does away with politics altogether. Life without politics is life governed by experts in harmony. If experts of different expertise differ – whether they have conflicting views or conflicting local interest – they may appeal to the super-expert. The super-expert has a traditional name all to himself; he is the philosopher king of the old myth of Plato's *Republic*. Some say Plato's *Republic* was but a dream, not a real plan, only a naive thought-experiment. Perhaps. Let us ignore Plato for a while, then, and move to the realistic image of technocracy.

The realistic picture shares with the naive picture the ideal of harmony and the means of scientific technology. It differs from the naive picture by taking notice of the fact that conflicts of interest and differences of opinion are bound to arise everywhere and be disastrous unless checked. It presents professions as organic unities run by organically grown leaderships. Politics will be definable as the technology of keeping harmony between the professions, and politics can be democratic or not. Technocracy requires the application of elitism: politics is for professional expert politicians, and this is precisely Plato's *Republic*.

A partially democratic image of technocratic society is conceivable. Michael Polanyi, one of the greatest philosophers of the mid-century whose reputation is justly on the increase, advocated a view of scientific-technological society as a conglomeration of professional guilds run by elites. His idea was popularized by Thomas S. Kuhn, one of the most popular philosophers of science today. Kuhn does not discuss the broad setting of this society of associated guilds. Polanyi does. He clearly is a democrat, and on the basis of one argument which seems quite sufficient for anyone democratically inclined, so that it is hard to assess its force. It is the view that unless society is democratic the government will interfere with the autonomy of the guilds.

The guilds in Polanyi's picture of society are research guilds. The guilds integrate in society via the educational institutions and via the provision of licenses for individuals to practice the specific crafts which the guilds purport to master. But all else, Polanyi thought, must be democratically run. Thus, also, the flow of individual recruits and of funds for the guild must be run democratically. Yet this clashes with the fact that these, too, are subject to expert handling.

Polanyi's picture of a democracy with guilds is untenable unless a line is drawn between the two. What subject-matter is open to all and belongs to democratic government and what calls for expert handling and thus calls for elitist administration? Did Polanyi himself deem religion, for example, not open to experts? Where is the line to be drawn?

A simple, attractive, and false solution is that short-term goals are best given to experts and long-term ones are matters for political debate and decision. This is like the proverbial wife who does all her sexist husband tells her to do, and agrees with his opinions implicitly, except that politically he is a conservative and she is a radical. Moreover, scientifically speaking, the technocrat should insist that both are dated: both conservative and radical politics should give way to the new political expertise. It is clear that in political life short-term decisions, like the annual budget, are vital to significant politics, yet such decisions must be limited by some long-term decisions.

Why, then, should we not have a philosopher king? The philosopher king, to repeat, will be unable to work out long-term projects (except survival, if we are lucky), and he will therefore look at the whole society through the utilitarian spectacles of increasing the efficiency of whatever is done most inefficiently. The irony is just here. Plato viewed this as a state of pigs and opposed it with the state run justly by the philosopher king, where justice is defined as elitist: the worthier gets more in proportion to his worth. When finally with the rise of technology elitism is becoming practicable in the form of technocracy, it turns out to be the state in which it becomes inescapable to live like robots. The title of F. G. Junger's book, *The Failure of Technology: Perfection without Purpose*, says it all; where the failure mentioned in the title is not of technology but of a society drifting towards technocracy. Democracy should try to pinpoint the failure, analyze it, and repair it.

What is the failure? It is of the view of science and technology as prescribing unquestionably optimal solutions to all pressing problems. This view is true as far as it goes, which is extremely important, but it is extremely limited since science has still not answered most of our questions, and scientific technology is still most narrow in its fields of application: most technological problems, particularly most administrative problems, are unsolved; those solved are often handled by traditional and/or ad hoc means; and their solutions are seldom unique.

Technology commands assent, and by law: we are not allowed to market any innovation, any drug, machine, or procedure, unless we have a license which is an approval, a sign of public endorsement. This, however, does not mean that an obligation is involved here – only license. Technocracy turns license into obligation.

Classical philosophy of science, and by extension classical philosophy of technology, viewed science as the peak of rationality, rationality as obligatory and rational conduct as the conduct in the light of reason. This bespeaks of a unanimity of obligatory rational conduct. In the eighteenth century, the myth prevailed that science is the province of every rational being, namely of every normal human being. This myth led to populism which is democratic in that it makes participation in government possible and even obligatory to all but which is illiberal to the extreme. This was described admirably well by Jacob Talmon in his *Origins of Totalitarian Democracy*, 1955, which justly and immediately turned into a classic. Interested readers are also advised to look at Richard Hofstadter's exciting studies of nineteenth century American populism.

All this should suffice as a refutation of classical philosophy of science. But rather than give up its weakest point, its identification of scientific rationality as obligation rather than license, philosophers gave up its strongest point, its identification of scientific rationality as free for all. They concluded that science is not given to all, only to experts. Thus their populism became elitism and their democracy was lost. And classical philosophy of science, which suffered from the democratic view of reason as obligatory, changed into an elitist view of reason as obligatory. There is a saving grace, though: democratic theory and democratic practice developed into liberal democracy, where earlier toleration and the right of dissent was entrenched.

It is high time to unify rationality with the ideas of toleration and the right of dissent: democracy and rationality should be united within a liberal philosophy of man. The first step in this direction is to distinguish between the freedoms of thought and of action: the freedom of thought should be unlimited and should rule science; the freedom of action must be limited, though its limitation should be as little as we learn to minimize it, and those in charge of it should be monitored and limited.

5. CONCLUDING REMARKS

Technocracy confuses ends with means. The expert travel guide is very useful, and can advise the traveller how to go to where the traveller wants to go, or where to go in order to find whatever it is that the traveller seeks. Now what the traveller seeks is his own affair – his own goal or his own means to achieve a distant goal; what the travel guide offers is a possible way to achieve a given, short-term goal. Remembering well this simple metaphor suffices as means for the prevention of the risk which a democratically-minded individual may face in our society, of uncritically slipping into a technocratic mode of thought.

This is not to say that political leadership is comparable to travel advice – merely that the individual citizen is as responsible for his political choice as for his choice of travel plans. When politicians propose national goals and priorities for democratic decisions, when they propose policies for democratic considerations, when they arbitrate, then they are neither technocrats nor travel guides but in between.

PART TWO

TECHNOLOGY AND PHILOSOPHY

INTRODUCTION TO PART TWO

This part presents a detailed picture of the classical philosophy of the Enlightenment. Since this philosophy is responsible for the magnificent growth of science and of scientific technology over the last three centuries, loyalty to science often seems to require the endorsement of that philosophy – hence its popularity, the dogmatic adherence to it and the political, social, and technological damage that this adherence is responsible for. It is important to notice in detail the intellectual strength and weakness of the traditional philosophy of science and technology: the good and the bad consequences it holds for science and for technology. It is desirable to seek an alternative to it which encompasses its intellectual and practical strength and overcomes its intellectual and practical weakness. It may be hoped that a better alternative will open the road to vigorous social and political technology, to the evolution of a more enlightened society, and to the development of efficient participatory democracy and effective democratic controls.

CHAPTER 4

MAGICAL AND SCIENTIFIC TECHNOLOGY

It is hard to say what anyone would find most specific, most characteristic and most typical, of one's own environment; yet it is customary to characterize Western modern society by the prevalence there of machinery and gadgetry in almost everyone's life. Therefore, it may be important to stress this: modern (scientific) technology is prevalent in only part of the world; technology is prevalent in all human societies, including the most primitive. Moreover, in most known societies technology equals magic. Since magic is so widely misunderstood, this equation is not as easily understood in the West as it might be.

1. FROM THE SCIENTIFIC POINT OF VIEW: BACONIANISM

What is known in modern society about magic? It is hard to say; people steeped in any religion that makes them read the Bible have one idea, engineers have a different idea, social anthropologists who devote many studies to it have yet another. To avoid entering the details of the disagreement, it is advisable to remain as superficial and uninformative as possible. It is too easy to describe all manner of magical rites, and too facile to cite scholars about magic; it may be more useful to offer a few comprehensive views of magic, and to quote as little and as few instances as possible.

For most people there is no difference between magic, witchcraft, and sorcery. So let us not distinguish these. Also astrology, clairvoyance, foreknowledge, or oracular powers can all be subsumed under magic, though here things get slightly complicated. Postponing the complication for a moment, we can say that for most people magic is just a set of superstitions, lumped under the label "mumbo-jumbo" or "hocus-pocus". The more knowledgeable identifies magic as the belief that certain words are allegedly potent: "open Sesame", allegedly opened an invisible door in a rock wall leading to a cave full of treasures. Yet in the story of Ali-Baba, things are relatively simple: anyone who knew the secret password "open Sesame", we are told, was free to enter the case. In this way the story of Ali-Baba is not a reliable guide to magic. We realize that magic stories are fundamentally

79

no more incredible than the latest achievements of science. In science, just as in the story of Ali-Baba, things are fairly straightforward: anyone can be taught to use a magic formula; yet in real life mumbo-jumbo is different from science.

The simplest idea is that unlike science, which is clear-cut, magic is confused and bound with all sorts of superstitions. It is not problematic that "open Sesame" does not work; nobody expects it to work. Mumbo-jumbo, whether love potions or the philosopher's stone or the fountain of youth or the ill luck caused by an evil eye, by a cracked mirror, by a turned-over salt dish, or by walking under a ladder – these are superstitions, i.e., beliefs of simple folk that are unexamined, and in confusion people ignore the fact that they are all simply false.

To take a fairly recent example, there was a folk belief that red-headed poeple do not suffer the ills of alcoholism, namely that they can safely indulge in excessive consumption of alcoholic beverages. This belief was deemed by one statistician popular enough to merit examination. Result: Negative. Query: Why did this belief remain popular for so long? Answer: Both alcoholism and red hair are sufficiently infrequent for it to remain unclear whether they are statistically correlated or not until the question was carefully examined. Query: why red hair, of all the rare characteristics? Answer: there is a superstition correlating red hair with all sorts of devilish qualities, such as hot blood and the disposition to flirt with danger and the ability to do so at a lower cost than the average. Query: is this superstition refuted too? Answer: No. Query: Why? Answer: it is too difficult to test it, since it is so vague, and so far no one has bothered.

Is this all there is to magic? Some say, yes. They are the Baconians, the followers of the great – perhaps the greatest – philosopher of the modern world, Sir Francis Bacon, who was a contemporary of William Shakespeare. Bacon said that magic is wrong not on account of the wonders it claims to be able to perform: science will one day outdo magic on this score; rather, magic is cranky, muddled, superstitious, and plainly false.

There are two obvious questions to ask the Baconians. First, can you equate all sets of superstitions and regard all that is not science as super-stition? Second, what is the position of the magician, then? Bacon's answer is very, very clear: all that is not science is both pseudo-science and super-stition. And all who advocate superstitions, be they magicians, alchemists, priests, professors, cabbalists or witch doctors (all doctors, in his opinion, really) – all of them are nothing but charlatans, petty crooks, and con men (i.e., confidence tricksters – people who exploit other people's credulity for their own benefit). In particular, religion should make no factual claim, and when it makes any factual claim, an astronomical claim, for example, or the claim that a place of pilgrimage is in possession of healing powers, then it is deception, no better than any other deception.

We can now return to a point left open a while ago: foreknowledge or prediction. We have two kinds of foreknowledge, scientific and magical – and, according to the Baconian view, none other. And there is one and only one difference for the Baconians between the two: scientific but not magical foreknowledge is reliable. As magic claims reliability, it is a pseudo-science. All talk about one oracle being more reliable than another is superstitious nonsense, and rooted in gullibility, and indicates the presence of people who pose as spokesmen for that oracle, usualy the priests at the temple of that oracle, who are petty criminals. It is the same – still according to the Baconians – with people who pose as mediums and as clairvoyants, whether in gypsy caravans or before the glaring floodlights of the television studio.

Here some readers may start feeling uncomfortable. The trouble with the Baconian theory is that it is too prosaic. It is hard to compare a crooked bookie to an oracle, and both to a prophet. Yes; where does the prophet enter the picture?

Here the Baconians split many ways. There are the Baconians who are religious themselves in the sense that they believe in some prophecies and in some miracles. (There are other senses of being or not being religious; they are of no concern for the present discussion.) They may make some allowances for the factual claims made by their own religion to the exclusion of the factual claims made by other religions. The ways they manage to do so are uninteresting, since their naïveté is open for all to see, and if they cling to their naive faith it is hardly possible to dissuade them. Alternatively, they may declare their religion's claims scientifically proven – whether by logic or by facts; it does not matter here. Their credulity is even bigger. Alternatively, they may deny that their religion makes any factual claims. This denial has a name: Euhemerism, after the ancient author who said that the myths of Greek religion, though rooted in true stories, are mere parable. Nowadays Euhemerism embraces the view that the bible stories, too, are parables. This newer view was proposed by some pious Christians, perhaps by Bacon himself, if he was a Christian at all, certainly by his follower Robert Boyle of the famed Boyle's law (the density of air in a given temperature is proportional to the pressure on it), who was a foundation member of the

Royal Society of London and its leading light. It is one thing to propose this view of Bible stories as mere parables as traditional, namely to distort history, and another thing as a new approach to the Bible. As a new approach this has yet another name: demythologization. The name has been popularized by the twentieth century theologian and philosopher Rudolf Bultmann. Finally, there are those Baconian writers (the others will be mentioned later) who waver between two or more of the views described above. The most famous American religious philosopher of the middle of the twentieth century is perhaps Charles Hartshorne, who confused all these. Thus, out of all the religious Baconians, namely, those who agree with Bacon about putting all non-science in one bag yet who accept their own religious denomination's factual claims, only the true demythologizers make any sense at all – only those who propose to strip their own religious denomination of all of its historic factual claims. Whether they make good sense is another question. Their view was once – since Boyle – held in high esteem. Today their religious peers and their scientific peers do not value them and together attack them with the question: why stick to one religious denomination rather than shift allegiance to another?

With this question left hanging in mid-air, we can now move to other Baconians. Some other Baconians are frankly anti-religious, and they consider their opposition to religion as a corollary to their Baconianism. The majority of Baconians, however, prefer to avoid expressing an opinion on the matter. It was, indeed, tabooed by the Royal Society of London to discuss religious matters in scientific circles. This is true of even the famous anti-religious scientist, Laplace, so famous for his answer to Napoleon's rebuke for his failure to mention God in his great book on the mechanics of the heavens: Sir, he retorted off-hand, I do not need that hypothesis. Also famous is the response of Lagrange, no less famous as a mathematician and as a theoretical physicist. When he heard that story he said, but what a beautiful hypothesis that is! What is so remarkable about this exchange is that it was so brief and so forced: neither the believer Lagrange nor the heretic Laplace liked to talk about God.

The taboo about discussing religion in scientific circles was political: politics was taboo, too, and religion was – and still is – a prime factor in national politics. And so science grew in a society full of magic and superstition – a most significant fact. The taboo was broken with the rise of Darwinism. Darwin tried hard to keep to the taboo, but he simply could not. His *Origins of the Species*, incredibly, avoids discussing the origins of the human species. Yet, he fooled no one; and, indeed, a few years later his

The Descent of Man followed. The first real blow to traditional Western religion was meanwhile dealt, and science came out with an open all-out assault on religion. The pretext was that a clergyman in Oxford attacked Darwinism in the name of the Bible. Supposedly he thereby forced science to respond to the allegations and so made the taboo impossible to keep any longer. This myth about one obscure clergyman who foolishly forced science to attack religion is breathtaking in its stupidity; the gullibility with which many clever people (including Darwin scholars) still accept it is really remarkable. Obviously, the myth is true but quite irrelevant: science violated Western religion with the Copernican revolution, since unquestionably the astronomy in the Bible, scanty as it is, suffices to tip the scale in favor of geocentrism as opposed to all other systems, heliocentrism or any other. And soon astronomers denied that the universe is as old as calculations based on the Bible would have us believe. Whether science openly attacks religion or not matters little, then; the result is simply this. The more we examine factual claims that religious people take from the Bible and compare them with science, the more obvious the conflict becomes, and, clearly, science is always the winner. This simplistic view is expressed even by the greatest philosopher of the twentieth century, Bertrand Russell.

2. FROM THE ANTHROPOLOGICAL POINT OF VIEW: FUNCTIONALISM

An older contemporary of Russell, Sir James Frazer, himself an enemy of traditional religion no less than Russell, and in many ways as staunch a Baconian as Russell, started a new trend of thinking, and precisely because he developed the Baconian view of magic as a speudo-scientific technology. (He discovered that Christianity incorporated older heathen superstitions and wanted to account for the failure of science to eradicate superstition, including Christianity, which he thought very poorly of.)

Let us return to prophecies and miracles. There is a fundamental difference between the foreknowledge of a prophet and that of science: the scientist and the scientific technologist predict on the strength of the laws of nature, whereas the prophet's power is supernatural. Query: Does the heathen's oracle resemble the prophet or the scientific technologist? Frazer answered unhestitatingly: The oracle is more like the scientific technologist in having a regular assignment rather than like the prophet who speaks if and when inspired; the oracle is like the prophet rather than like the scientific technologist in being unreliable. The oracle is, thus, a pseudo-scientific technologist. This, remarkably, makes the weather forecaster, since he is unreliable, a magician, rather than a prophet. The same holds for the regular Kremlinologist. Or take the miracle of the resurrection – of the child whom the prophet Elisha resurrected, of Lazarus whom Jesus resurrected, or of Jesus himself. These miracles, being one-timers, are different from the miracles at Lourdes which happen repeatedly with each fresh wave of pilgrims. Lourdes' healing power is magical; that of Jesus is not (unless we view Him as a systematic healer).

Perhaps all this is a great oversimplification of the view of Frazer: it is certainly an over-simplification of the facts of the matter. For, the very question posed – is magic natural or supernatural – is already an oversimplification and a trap. The very distinction between the natural and the supernatural is a great discovery, which magicians are (and prefer to remain) ignorant of; so why should they be able to see their own view as naturalist or as supernaturalist? And if they could not see themselves as either, can we do so without thereby distorting the facts?

The great twentieth century Hinduist philosopher, Erwin Schrödinger, of the famed Schrödinger equation for the electron, wrote an exciting short book (all his books are short, profound, and beautiful), *Nature and the Greeks*, in which he explains the greatness of the great Greek discovery of naturalism. Without doing justice to his work we may notice, what Sir James Frazer already noted at very great length (his magnum opus is a classic book about magic, twelve volumes long, called *The Golden Bough*), that all prescientific thought except for traditional religion is both animistic and magical. Animism is the ascription of souls to natural objects, whether trees, brooks, mountains, or even meadows. Souls are in possession of intentions, good or bad. Primitive people (whatever this expression exactly means – but let us leave this for now) perceive objects and even forces of nature as being for or against this or that person. This perception seems to be a corollary to animism, a perception which makes animism go naturally with magic. Even religion has a touch of this perception in the form of anthropocentrism – the doctrine that God has created the whole of Creation to serve Man. Anthropocentrism is one point where Aristotle and the Bible agree. But neither presents the forces of nature as friendly or as hostile. (This is not quite correct: there are some exceptions. The most important exception is the rainbow that the Bible presents as always friendly, and to the whole of mankind. The Baconian naturalists Descartes and Spinoza pressed home hard the claim that the rainbow reflects a law of nature since – notice this "since" – this claim renders the rainbow neither friendly nor benign.)

Animism gives magic its inner logic, its convincing force. Why should a piece of rock move when I utter the sound "open Sesame"? A stone has no understanding! A naturalist will say that a machine is put in the stone which responds to this sound pattern with no regard for its meaning; a few science fiction stories (and one *Star Trek* episode) follow this naturalistic idea. A Magician will say that though the stone has no understanding, the spirit which resides in it does, and may very well obey!

Does the spirit of the rock have to obey the command "open Sesame" every time the command is uttered? Does it have no choice? Does it matter whether the person who utters the words is a friend or a foe? These are dangerous questions. If the answer is yes, then we have here a (putative) law of nature; if the answer is no, then perhaps we have here a supernatural force. Hence, the question is better unanswered, or, still better, unasked.[2] And if asked, it can be evaded, or we may say that it depends on how the spirit reacts. If it always reacts favorably to strongly convinced people, then it is still natural; if it always reacts favorably to holy men and women, then, perhaps not.

Any analysis of magic will prove it ambiguous on this very point, regardless of whether the magic is Eastern or Western, ancient or modern. We may, therefore, claim that magic has nothing to do with technology, that technology is purely naturalistic. Did Frazer agree? In a sense, yes; but only in a sense. This brings us nearer to the view of magic as a technology, regardless of it being animistic, or rather, regardless of whether animism can or cannot be viewed as naturalistic. Let us take this point slowly.

Frazer noticed that the vagueness of magic claims is elastic. Bronislaw Malinowski pressed this point harder, yet to a different end: seeing its worthlessness as technology, he sought its worth elsewhere. He found it in the explanation schema that Emile Durkheim had offered. As theories, religious and magical doctrines looked to both Frazer and Malinowski as senseless as they should look to any Baconian. Frazer tried to explain the prevalence of senseless views while taking them to be, on the whole, views; following Durkheim, Malinowski objected: he took them to be symbolic expressions of their practitioners' awareness of their conditions, especially of their need for social cohesion and integration. Durkheim saw primitive society as less organically knit than developed society, and magic as less cohesive or integrative than religion. Malinowski disagreed and presented magic, too, as highly integrative. People go to health spas for their ailments and complaints. There is no doubt about the curative value of the water offered by most spas for ailments: it is nil. We can condemn going to spas, drinking a lot of water and spending a lot of money, as nothing but superstition and folly. Or we can take it as the excuse of the rich to go on holidays, etc., etc. The more sophisticated our view of the reason and function of going to health spas, the less the question signifies, do the sick, imaginary or real, believe in the curative value of the waters? Take another example: camping. People hardly offer a reason for their preference of rough conditions for a few days a year, and sociologists might offer a few explanations for their conduct. In some countries camping is an expression of national unity, especially among the young; in other countries it may be an expression of rebellion against the national ideology, especially among the young. Any view correlating camping with national unity that campers may offer, which seems to be a factual claim, will have to be viewed as scientific or superstitious – or as symbolic expression of the significance of the act it accompanies. And so, for Malinowski the symbolic value of magic and of religion, as well as their integrative values, are variable from case to case.

Most societies we know are fairly primitive, especially technologically: most societies are pre-industrial and based on hunting-and-gathering (foraging) or on agriculture. Most magic rites we know are personal or social. The personal ones have to do with rites of passage – birth, initiation, marriage, sickness, death – and with personal fortunes and misfortunes. The social and the socially-oriented magic relates to the economy, to a specific hunt, a war, or to the acts of sowing and planting. Magic rites relating to these acts abound. These rites are, indeed, part and parcel of these acts and accompany a significant part of the technology they involve. Do the rites matter? No: they resemble the breaking of a bottle on a new ship, a house-warming party, and other rituals that are common in the technologically sophisticated Western world.

The picture depicted in the previous paragraph must be deemed suspect: what has happened to animism? It has dropped out of the picture altogether. The reason is simple: animism is false, even confused, and so to ascribe it to most of mankind is to put our own civilization in the right and the rest of the world in the wrong. This seems highly unpalatable (I. C. Jarvie, *The Revolution in Anthropology*).

And so Baconianism has undergone a total reversal. When Bacon started his grand scheme he said that there is hardly a sane man around; everyone is so steeped in prejudice and superstition. He taught people to reject out of hand all they had learned – not to refute all they had learned, since this consuming and hardly worth the effort, but simply to reject all past learning out of hand. The reader may remember the superstitions about red-heads and how hard it was to test – and refute – even one of them. This may amply justify Bacon's demand. Anyway, the learned world accepted it – at least the avant-garde, at least concerning things natural though not always concerning things supernatural. Well, perhaps no one ever accepted Bacon's demand; perhaps it is even impossible to do so. Nevertheless, the historical fact remains: Bacon's view became increasingly popular and its converts became philosophers, as they were called, and invested their efforts in experiment and in technology.

They made the West scientific and technological. Perhaps Bacon – and his followers Descartes, Spinoza, Locke, Newton – had no immediate influence on the growth of the modern technological world; yet their indirect influence is what made all the difference. At least Baconians should think so. And Durkheim and Malinowski were Baconian indeed. Yet they refused to grant the intellectual superiority which the West usually claims over the rest of the world. This is quite an about-turn, and a very disconcerting one. It should be particularly disconcerting for Baconians, but even non-Baconians must object. It is at least puzzling, to say the least, that science militant claimed superiority of the wildest kind, yet science triumphant rejects all claims for superiority as all too parochial.

The Durkheim–Malinowski theory is a generalized Euhemerism. It declares all prevalent superstition to be a social expression and ignores all prevalent superstition as beliefs proper. It thus applies, as Durkheim has stressed, to all religions, ours and theirs. Yet we do know that scientists do not believe in health spas and so are less likely to go there, no matter what social significance spas have, and they are even more unlikely to go to Lourdes. Why should we take a rational view as an expression of a belief and an irrational view as merely socially significant ritual? Does it make all men equally? Does this not contradict the claim that scientists shun Lourdes? What we cannot deny is that science is very unusual. What makes science so very special?

3. FROM THE METAPHYSICAL POINT OF VIEW: FIDEISM

Enter the anti-Baconians. Most important among them are the (creator of Roman-British archeology and) philosopher R. G. Collingwood, the (Malinowskian) social anthropologist Sir Edward Evans-Pritchard, and the (natural and social scientist and) philosopher Michael Polanyi. Every view, they said, signifies only within an intellectual framework. An intellectual framework is a set of absolute presuppositions, unquestioned and taken upon faith. Science, too, has its presuppositions. These may change, but must be taken upon faith at some stage nonetheless. Hence Bacon's counsel to do away with all preconceived opinion is even for science quite impossible. Moreover, since the validity of each assumption depends on the framework within which it is placed, each framework is utterly isolated from others: they may well condemn each other, but to accept a condemnation of one given framework is only possible after the acceptance of some other framework, one which is equally condemned by the one given.

What has been presented thus far is a modern variant of a traditional religious philosophy known as fideism, where the stem "fide" (as in "confide", "confidence", and "fidelity") means faith. Fideists agree with skeptics that we may doubt every assumption, no matter how well proven, because we may also doubt any method of proof and any criterion of proof that may be used. Unlike the skeptics, however, the fideists refuse to doubt every supposition, because doubt paralyzes and yet we must act, or else we die.

Most twentieth-century philosophers are fideists. Usually, since fideism is based on the pragmatic argument that we need faith for practical purposes, fideists tend to be pragmatists. They recommend those presuppositions that are most conducive to survival, namely the presuppositions of science. What are these? We do not quite know, and it does not much matter anyway, if we are willing to endorse science as a whole, lock, stock and barrel – presuppositions, theories, and anything else which happens to go with them. Nevertheless, one thing can be said with little hesitance: one presupposition of science is naturalism, a doctrine we have already met: our experience is explicable by means of laws of nature and so nature is governed by laws and not by intentions.

This is another about-turn. At about the same time in which Baconian sociologists and social anthropologists have tried hard to give up the Baconian idea that magic is superstitious, Baconian philosophers gave up the Baconian idea that science is the only rational product of the human intellect. Instead, they accepted science on faith – on the same ground on which so many different (religious) systems were accepted and advocated in the course of human history. Even naturalism, the mainstay of the scientific attitude, is now taken upon faith alone, on the pragmatic ground that modern scientific technology is the most powerful! There is irony here: pragmatism goes better with animism, with judging everything as good or as bad for us; yet science is good for us and is anti-animistic. Hence pragmatism is an error; yet pragmatists merely conclude in their animistic fashion that anti-animistic science is good for us and so is better judged right!

Does faith in technology imply faith in naturalism, and if so, is not fideism now bound to impose science? This is a tough question. In the South Seas there are new magical or religious sects that come to life repeatedly with a disturbing persistence and which are known by their magical or ritual practices or cults, usually called cargo-cults. (The accent on ritual is due to the anthropologists who report them, not due to the facts of the matter. See I. C. Jarvie, *The Revolution in Anthropology*.) A cargo-cult ritual is designed to bring about modern technology. What it is matters little; it may be the construction of airfields or of wooden toy airplanes, or any other magic ritual. Sir Francis Bacon suggested that experimentation is a ritual meant to force Mother Nature to reveal Her secrets. Perhaps naturalism is not a philosophy but a magical animistic nature worship! This idea has been proposed by a leading twentieth century biologist, Sir Charles Sherrington, in the opening of his classic *Man on His Nature*.

The social anthropologist Sir Edward Evans-Pritchard argued that magic is an intellectual framework as legitimate as science. It is a false claim that declares the magically minded gullible and credulous, he reports, Magically minded people, for example, test oracles. It is the idea that all oracles are wrong that the magically minded cannot understand: it is a part of his intellectual system that some magic is correct. Moreover, he can and often does explain facts that science cannot explain. For science coincidences are seldom explicable; and when they are explained they are taken to be results of other, inexplicable coincidences. Generally science leaves coincidences as inexplicable – as sheer accidents, as meaningless. Magically minded people cannot abide by that: they find meanings in all coincidences – usually good or bad intentions (whether of someone or in the abstract).

(Consider Ali-Baba who by sheer chance overheard the robber chief say the magic word "open Sesame!" Evans-Pritchard should view this story as a poor account of magic.)

One who agrees that an evil eye may spoil a crop will have to admit that fending off an evil eye is as much technologically important as the fending off of all sorts of pests or even as the sowing of seeds. The question is, is the evil eye effective and will this or that ritual effectively fend it off? This is a very hard question. So is the question, will this or that medication help or worsen matters? Very often missionaries found it impossible to explain to natives events that have run contrary to their expectations. The Baconians love the story of a traveller – was it Captain Cook himself, or only Alan Quatermain, the brainchild of Rider Haggard? – who saved his skin by predicting a lunar eclipse. The facts are more complex. Who can tell that there is no evil eye, that there is no hidden reason for the technological superiority of some nations over others? Surely, the technologically superior have a scientific explanation for this fact. But the magically superior have a different explanation. Should technological superiority come with scientific explanation? Is it necessary for high technology to come with science, naturalism, and all that? If so, is it worth it?

Dr. Sun Yat-sen, the father of modern China, had hoped otherwise. He found his culture superior to European culture in every respect except the technological respect. He had hoped to maintain Chinese culture while adopting Western high technology, especially high military technology. He failed. Could he have succeeded? Was it sheer accident that he failed? Or an evil eye? Or was he doomed to fail? If he was doomed to fail, should China have preferred the superior Confucian culture over guns or the other way around? How can one decide such matters?

Western scholars are fairly widely agreed that the Chinese, as well as the Indian and the Japanese, exhibit high cultures different from Western culture. The main concern of an Oriental cultivated man is personal salvation, be it of the Hindu type, the Taoist type, or any other. This makes the mystic selfish, as Max Weber claimed, when he characterized the mystic's desired end as peace of mind and the mystic techniques as the means to that end. Does this not make mystic exercises a kind of technique that must be noted by students of technology?

Usually the word "technology" is applied to physical engineering, at times to biological technology, especially medicine and agriculture, hardly ever to other fields such as education or psychoanalysis or behavior therapy. Yet there is no reason for this other than certain Baconian prejudices which we'd better examine rather than endorse dogmatically. And if we ever agree to include under the heading of technology any kind of human technique, educational, organizational, or psychological, then we shall have to include Yoga exercises too. Of course, one may object to human technology in general as immoral since it manipulates people. Of course, one may object to Yoga as mumbo-jumbo. All this is a legitimate topic for a separate and subsequent discussion. First, however, it should be decided whether military technology and Yoga exercises can in principle go together as technologies.

How should one answer this question? Consider the following questions. (1) Has there been a yogi who waged a war? This question leads to controversial answers. But, much worse, it is a controversial question: one may wish to replace it with another question as more suitable to the present context, one that pertains to the culture which produces yogis or Buddhist monks or Taoist monks. Let us, then, observe the following question. (2) Does such a culture also produce a war technology? This question has an uncontroverted answer, since everyone agrees about the following general sociological fact: there is no culture which has no war technology to it. Hence, when we wish to move from the individual to the social, we should not replace question (1) by question (2), but perhaps by the following question: (3) Can the culture that has produced Yoga technology continue to produce it while creating high war technology (including nuclear weapons)?

At first blush this seems a silly question. Why should the yogi bother with guns when he seeks peace of mind any more than with bows and arrows? But this is a misapprehension. The culture that has produced Yoga also engages in war. Perhaps the scientific culture that goes with guns and nuclear weapons ousts the culture that goes with mumbo-jumbo, and perhaps Yoga is a lot of mumbo-jumbo; and then importing guns and big bombs will oust Yoga. Is this so?

This is the classical Baconian way of presenting matters – as above but with no perhaps: Baconianism is fairly dogmatic in its contrast between science and dogma. The Baconians of the fideist type will probably agree: they do not accept Baconianism as self-evident but as having high survival value, yet they will admit that science and mumbo-jumbo are incompatible, as a matter of observed facts. Even the anti-Baconian fideists might agree on this matter, even while stressing that science and Hinduism are two separate systems and as such incommensurable, i.e., as not given to judgment in a superior court, as judged (negatively) only in the courts of law of each other. Some anti-Baconian fideists, particularly Sir Edward Evans-Pritchards and Michael Polanyi, would be of a different opinion. They would advocate that one and the same person should be both a scientist and a Roman Catholic (even though Polanyi was a Jew and a Methodist and Evans-Pritchard an Anglo-Catholic), and point at the obvious fact that some of the world's greatest scientists and philosophers were religious people and even religious Roman Catholics.

All these schools are so obviously in error that one may wonder at the enormous capital that is invested in so trite a debate. It is one thing to be a practicing Roman Catholic and quite another thing to go to Lourdes to get cured of cancer. And it is a fact that the better informed and educated a person is – better in the light of modern science – the less likely that person is to go to Lourdes for the treatment of a physical ailment. The question, then, is this: are Yoga practices of necessity combined with mumbo-jumbo?

This is a very tough question, as anyone knows who is even vaguely familiar with the life of Mahatma Gandhi, the yogi most famous in the Western world to date. We can sidestep it with ease, however: the way Yoga is practiced in India today is, as a matter of fact, a part of a complex Mumbo-jumbo system. Can it be purified? Or perhaps it must remain superstitious in order to survive?

4. FROM A HISTORICAL POINT OF VIEW: THE SCIENTIFIC REVOLUTION

Malinowski-style social anthropologists are super-Baconian in their taking for granted that one part of the ritual of a farmer is scientifically attested and the other is scientifically pointless and so must gain its worth from its symbolic meaning. In this they are mistaken, since the farmer performs one complex operation and cannot divide his activity into two parts, and distinguish the attested by the scientist from the symbolic; much less can he give up one part of it as merely symbolic. After he develops a scientific conscience he may be able to do so. He may then go on performing the symbolic part of his practice as merely a symbolic act. He will then be acting like the person who knows there is no point in breaking a bottle of wine when launching a new ship, yet who does it in good humor[3].

Bacon and the whole Enlightenment movement demanded the radical renunciation of all errors as superstitions, claiming this to be the first step towards the evolution of science proper. In our own historically minded century an increasing number of philosophers view science as an outgrowth of superstition; even while seeing a fundamental difference between science and superstition, they deny that either may be found in pure form. And this move transcends Baconian radicalism, Baconian fideism, and anti-Baconian fideism: all three are ideal images of science, not images of real science in history.

This is an application of the theory of Claude Lévi-Strauss of primitive thinking to the problem at hand. Lévi-Strauss says that myths – which is the name he gives to all sorts of primitive peoples' superstitions, whether they come as stories or as general statements, whether they concern religion or not – myths come in opposing pairs are are used in mixtures of pairs, only mixed in varying degrees. That is to say, a pure myth is false, and so is its pure opposite: the truth is somewhere in between, and, moreover, for different cases in different distances from either extreme. A scientist would ask to what measures are two opposite myths mixed, and under what conditions do the proportions vary? A savage would not ask but (like a dexterous cook) mix his myths in varying proportions, depending on circumstances. This, we can see at once, makes his thinking unscientific and uncritical. To apply this to our case, Baconianism is the myth of pure science, (pragmatist) fideism is the myth of anti-science; science is somewhere in between.

Once we reject Baconianism, we also reject the classical Enlightenment view of society and of the place of technology in it. How to replace it with a view of technology more in tune with the twentieth century historical approach is the problem of the present study.

The Baconian view of the rise of modern science presents a watershed between the medieval and the modern, with the modern being the scientific revolution announced with the foundation of the Royal Society of London in the mid-seventeenth century, which has instituted the Baconian ideology. Before that revolution there were some forerunners, we are told, especially the Copernicans in science and the Baconians in philosophy, perhaps even some forerunners to Bacon himself; but by and large the population was ignorant and the learned were the mumbo-jumbo practitioners.

The current critique of radicalism is historical: every thinker has forerunners to whom he is indebted. The Baconians dismissed the claim that Bacon's forerunners were important contributors to the rise of Baconianism – since all forerunners did not know how dangerous superstition is and how important it is for a scientist to keep his mind pure.

This critique of radicalism, when taken to the opposite extreme (Pierre Duhem), leads to a highly conservative view of the growth of knowledge which permits no revolutions, only a continuous growth.

The truth is in between. Moreover, the debt we owe to our predecessors, we owe to the magicians among them. In the Renaissance prior to the scientific revolution there were the university professors who were church leaders, Aristotelian, rationalist, conservative, and the alchemists, astrologers and physicians, who were anti-academic magically-minded fideists, irrationalists who yearned for child-like simple faith, and whose faith in simplicity was magical. Both parties agreed that wisdom and knowledge were present in ancient books, except that the meaning of their message was lost. The tradition of search for the meaning of ancient messages evolved both in the casuistic university and by fideistic heretics: as in magic proper, the heretics believed that the secret could only be revealed to a holy man. They tried to be holy and deserving of the great deed of discovering the key to ancient secret knowledge. Thus, the heretics were anti-rationalist mystics, but unlike the oriental mystics who sought their own peace of mind (and condemned as selfish by Max Weber and by generations of Western thinkers), the Western mystics were practically oriented and tried to heal the sick as well as to revive the ancient Golden Age of science and religion. They tried to unlock the secrets hidden in ancient books by studying numerical values of letters, hidden cryptograms decipherable by combining and recombining words and letters in ancient texts. They applied ancient Greek mystic doctrines to biblical texts. Some of them were cabbalists, some astrologers, some alchemists, some physicians – but distinguishing between these groups is very questionable.[4]

The herald of modern philosophy of science was Giovanni Pico della Mirandola, a mystic who died very young: his call for renovation is to be found is his celebrated *Oratio on the Dignity of Man*, perhaps because of its very title and its contrast with the medieval view of man as a mere worm. (Remember Lévi-Strauss: the universe was created for Man; Man is but a worm. What a contrast!) He said very little, since he died very young. He was a nobleman who expressly viewed himself as a Christian cabbalist. Making him a kind of saint was the legitimization of cabbalah. In his *Oratio* Mirandola reaffirms his opposition to magic, since Holy Writ condemns it; nevertheless, he declares some magic to be good. The word for good magic soon became "natural magic", or, we would say today, scientific technology.

Mirandola's chief disciple was Reuchlin, who changed the name of their activity from cabbalah to Pythagoreanism, since numerology and the search for perfect proportion goes back to Pythagoras who was, as everybody knows, a personal disciple of Moses, from whom he learned all the secrets of Creation soon after the latter came down from Mount Sinai.

Around the year 1600 the leading thinkers – Bruno, Kepler, Gilbert, Galielo, Harvey – were all Pythagoreans, none of them in a university or in any other Church organization, though Bruno and Galileo had been (as a monk and as an academic, respectively) but left. They all believed in Copernicanism, since it is simplicity itself compared with its predecessor in astronomy, and since it put the sun, the symbol of light, kingship, gold, God the Father, etc., in the center where it obviously deserves to be. They all believed in mathematics, in proportions, in the science of combination known as mechanics. They all hated Aristotelianism quite passionately.

Yet there was a difference. In the opening of Galileo's first great work, his dialogue on the two great world systems, the Aristotelian interlocutor begins with an attack. To put it in modern English (the original is beautiful and highly recommendable), the Aristotelian says: you Pythagoreans condemn us as saying a lot of incomprehensible mumbo-jumbo, yet you are worse at that. And Galileo's mouthpiece does not deny the charge; rather, he – Galileo – promises to behave himself and talk clearly and simply.

And so he did.

Can we say that therefore Galileo was no longer a Pythagorean? Perhaps. The Inquisition addressed him, both in its accusation and in its verdict, as a Pythagorean. But they need not be followed: after all they were hired legal criminals. The Grand Inquisitor, St. Robert Cardinal Bellarmine, who instigated the legal murder of Bruno (who, in 1600, was burned at the stake) and who over a decade later threatened Galileo – he did not say with what – had a quarrel with Galileo over the true deep meaning of Psalm 19, the one about the sky telling the glory of the Lord, and the sun, and the stars in their courses. Psalm 19 is the text for Giovanni Pico della Mirandola's *Oratio on the Dignity of Man*.

Coincidence?

5. CONCLUDING REMARKS

What has been gained by a sketch of the different views of magic? First and foremost, that it is an unscientific technology, yet not mumbo-jumbo but a mixture of more or less valid information, items we would endorse, fused or confused with items we reject as all sorts of superstition. These are mainly views about good and evil forces that are not exactly natural and not exactly supernatural but which control and steer the course of events and cause seeming coincidences. The claims of magic are largely factual but not clearly so and clearly not always so, and even when clearly so, not in a sufficiently clear-cut and simple manner to be easily testable. In particular, according to magical views the powers of magic may cause some coincidences but other coincidences may invalidate these powers. Which way things move depend on the degree of potency of a given piece of magic. Yes, in addition, magic is potent when performed by a sufficiently potent and/or pure-hearted magician. The power of the magician, in turn, may vary, depending on some external powers. (Readers of comics should be familiar with all this.)

There is no way of telling beforehand whether a magical claim when translated into scientific language will stand up or not. It is not clear how to effect a translation. This point is of hardly any pragmatic significance these days, but it gives us an insight, perhaps, into the prevalence of magical technology and the peculiarity of scientific technology. It is useful to remember that traditional medicine is murky enough to resemble magic, whereas preventive medicine and magic medications – chiefly antibiotics – make many claims of medicine as clear as to render them scientific. Above all, contemplating magic may shake us out of our complaisant blanket assent to science. For, assent to magic is not different from the blanket assent to science. Science, in other words, or more precisely, scientific technology, is the magic of the simple-minded Westerner.

CHAPTER 5

THE IDEAL OF RATIONAL MAN

Baconianism has undergone many transformations and great developments. In many universities an advanced course is taught in that subject and is usually known as inductive logic or inductive policy or the philosophical foundations of probability or philosophy of science. The same subject is often taught in some departments of mathematics or departments of mathematical economics as games theory or as decision theory. Elementary economic theory also covers parts of the same ground, whether under the theory of consumers' preferences or behavior or even under the classical or the neo-classical theory of market mechanisms. In "hard" psychology departments the subject appears as learning theory or inductive policy. The whole complex of studies hardly ever comes into focus, although some attempts in this direction were made and are usually presented either as part and parcel of welfare economics – the theory of Pareto optimum – or as part and parcel of the critique of economics launched by certain sociologists and usually known as the critique of economic man, i.e., of the allegedly perfectly rational man or even of individualism in general. The philosopher Ernest Gellner has recently dismissed today's Baconianism as a combination of third-rate mathematics and fifth-rate sociology. Our concern with it here is as a philosophy, the very same radicalist philosophy already presented as Baconian and which Gellner presents as Cartesian (not that there is any quarrel here, since the radical Descartes was also a disciple of Bacon).

Gellner rightly admires the classical radicalism of the seventeenth century. He says that it has brought about what he calls the great divide – the great difference between scientific-technological society and the rest of the world. Yet he dismisses its expression in the twentieth century. Under what conditions is it useful to elaborate on the Baconian–Cartesian ideal?

1. THE BACONIAN–CARTESIAN IDEAL

Different kinds of criticism are levelled against the neo-Baconian image of the rational man. One of them is that this neo-Baconian image of the rational man is not true but a mere idealization at best. Neo-Baconians love this kind of criticism and even develop standard examples of it. The one most popular right now is the criticism from experience as reported by learning psychologists (Tversky and Kahnemann) about people's assessments of proababilities: people often bungle simple estimates: they bungle simple calculations which require a little patience to get right; they often bet on horses without the slightest concern with probabilities, odds, and all that.

When voicing such criticism, one seems to be very critical of the neo-Baconians, while accepting what constitutes their main claim. Their claim to have captured correctly the image of rational man is criticized and thereby their claim to have captured correctly the ideal of rationality is conceded. Naturally, they are very willing to accept this as a package deal. Moreover, it is hardly credible, yet a fact, that the neo-Baconians can easily handle the critical part of the package. Some sociologists endorse Max Weber's theory of ideal type, yet are very disgruntled about the neo-Baconian ideal type of the rational man just because it is ideal rather than factual. And this, of course, is not very fair: if Weberian idealization is accepted, then by the same token the neo-Baconian idealization may be accepted too. Moreover, accepting the Weberian idealizations, we simply must accept the neo-Baconian one: if the Weberian ones are legitimate, then perhaps neo-Baconian rationality is thereby also legitimized. For, what is the neo-Baconian idealization if not the one that is present in every Weberian one?

When Max Weber recommended that social scientists describe ideal types, whether of bureaucrats or business men, he did not mean to recommend that social scientists violate the facts. He simply felt that getting the facts roughly correct is already a great achievement, though still in need of correction. Quite intuitively, we feel that he is right, that there are characteristics of bureaucracy that transcend detailed differences between bureaucracies of different organizations or countries. And, of course, the main point of Weber's proposal is that we try to get the major and more universal characteristics right before we go into details: this is simply a matter of a sense of proportion. Yet, still quite intuitively, we may also feel that we cannot study the characteristics of all bureaucrats before we study the characteristics of some given groups of them, so that perhaps it is much more reasonable to be less ambitious and study the more detailed particulars first and generalize only in small steps and with due caution and after much detailed work has been assembled, studied, and compared.

It may be advisable to decide whether it is better to go into a general critique or into a detailed critique of the neo-Baconian theory of rational man, and this should be decided before we enter the detail of the critique. The neo-Baconians always prefer to start with the detail. And, indeed, they are enormously flexible and open-minded about details. Their dogmatism concerns only the general. Perhaps their worst general dogma is this: quite generally, they declare, it is better to start with details than with generalities: the other way around, they say, leads to dogmatism.

The detailed criticism one can level against the neo-Baconians is not only that their idealization shaves off the crude facts and renders the picture over-polished. The criticism may be from the most general considerations possible – from mathematics and even from logic. Usually neo-Baconians consider the detailed criticisms, and one by one, and with all the seriousness they may deserve. This keeps their minds off of the more general and more serious criticism for as long as they can work productively perfecting the picture by considering this or that detail.

One may present neo-Baconianism in the crudest way possible; it would then seem to some very stimulating and interesting and to others as pointless caricature. This is an empirical fact. A writer who meets such a divided audience response to a description may prefer to discuss it differently with the different parts of his audience in accord with their response. But in order to be able to do that he needs some market research: he needs to know what else characterizes the different groups than their different responses, and how to reach the one audience with the one sort of publication and the other with the other. This is easier said than done. Moreover, the same audience may feel under the pressure of criticism that they should change not the theory but its presentation. So let us first look at the crude presentation of the neo-Baconian ideal of the rational man, be it taken by the reader to be a caricature or a serious proposal. (Another presentation will be taken up later on.)

What characterizes rational man is his rational choice of hypotheses to believe in as well as to act upon when attempting to reach his own aims. He has, therefore, a set of preferences, a set of alternative hypotheses to choose from, a set of facts, and criteria of choice to apply to the above three sets. To be more precise, there are many sets of triplets of sets (preferences, hypotheses, facts) and a choice function which selects one hypothesis from among the competing alternatives within the second set of any given triplet of sets. One may put things differently and describe also a fourth set, a set of recommendations of options for action, namely what are the alternative courses open to the actor. There is no need for that, however, since the options for actions are described by facts and hypotheses; one hypothesis may say that administering a love potion will make the administrator of the option desirable to its consumer, another will say that it will make no difference, and yet another will claim the very opposite of the first. And so on. A fifth set can be postulated as well, a mathematical system of probabilities, likelihoods, utilities, and so on. This, too, is taken as a matter of course, and may either be absorbed by the previously mentioned sets or taken for granted as if already mentioned. When discussing presentation, much depends on custom: some scientists do not mention the mathematical apparatus they use as part of the presupposition of their theory, others do. One way or another, this is a mere presentational matter.

Most, but not all, of the mathematics that goes into what has been thus far mentioned is not very advanced and not very difficult. Nevertheless, it easily gets very complicated and very, very problematic. Consider preferences. Adam Smith already worried about the fact that we usually prefer a diamond to a glass of water, and quite unhesitatingly; yet when dying of thirst in the desert we will reverse the preference equally unhesitatingly. It is hard to see how to offer a set of preferences in the abstract when in reality they are so context dependent. But why does it matter? Why not add a set of contexts to the above set of sets on which choice is performed and be done with? The answer is that contexts are facts, and we want facts and preferences put in different sets so as not to confuse issues. (Neo-Baconians regularly assume that whatever they find indispensable is thereby proven possible rather than that it refutes their philosophy.)

A solution to the problem has been found. To be precise, a series of solutions have been found to a series of mounting problems. Scholars differ about how successful the current set of solutions is. The best known solution, in particular, is the theory of indifference curves; it is not clear from the literature how satisfactory this solution is because there is no comprehensive picture of its integration into the neo-Baconian ideal of rational man. In particular, the theory of indifference curves postulates the existence of an order among all possible states of the world: in any given moment each two are equally desirable to any given individual; either he prefers the one, or he prefers the other. This assumption looks very logical and very neat, especially since so much can be done with it in the theory of consumers' behavior without postulating a metric, that is to say without deciding how much more desirable is state a than state b, or state b than state c. For, in other parts of the theory, use is made of a metric, the so-called utility measure: a utility measure is the general answer (for a given individual) to exactly such questions. The reason a utility measure is introduced is that when banking on a probable outcome of an action much depends on how desirable it is: a very lucrative rare option, to take the standard example, may well be preferable to a more probable but less lucrative one. In this example money is used as a utility measure, yet we know that the utility value of money is high for the poorest people who in their poverty may starve to death or freeze in the cold, but perhaps very low for a person so rich that he cannot count his money or even assess his wealth. It is hardly credible, but arguments for socialism were made from the claim that it (socialism) is the only way to equalize wealth and the claim that this would maximize the all-around utility of money (since the utility of money always decreases.) Fortunately, for socialists and for their opponents alike, that episode is over. One may complain that many debates in neo-classical economic theory are still rather silly, but this matters little if the expected utility of an important debate is so high that its infrequency among debates does not deprive them all of the highest preferability.

Probability theory is supposed to make the rational man choose a hypothesis in the light of evidence and the chosen hypothesis imposes predictions on him and the predictions narrow his area of choice. If so, then hypotheses are eliminable, and may be viewed as merely mathematical ancillaries (and, it turns out, not even that): facts are projected with the aid of the probability calculus to forecast future facts, etc. As David Hume has argued over two centuries ago, this idea does not work: past facts never decide future facts, no matter what calculus one uses. And if one does have a projective calculus – a calculus to project past facts into the future – then why should one use it? Can there be alternative projective calculi? If so, how can we choose between these? Moreover, the neo-Baconians all take for granted all reports of observations of facts as if they are unproblematic. In fact, even the best reports of the best observed facts are often very problematic. How do neo-Baconians overcome these criticisms? By a simple fideist argument: we must act or die, and acting rationally is conducive to survival; hence, we'd better act rationally.

This is far from deciding issues, since the issue is, what is rational? For a concrete example, we can ask, given the facts, the options, the preferences, and whatever else that is required, should one prefer to act or to deliberate some more, perhaps also to seek more facts in order to be in a better position for proper deliberation in order to act? This is a beautiful question. As long as the mathematics for handling it was wanting, it was suppressed by the fideist argument just mentioned. Now that Abraham Wald and his followers have developed decision theory, his answer is triumphantly given to all Doubting Thomases. The answer is in principle frightfully obvious: the choice between acting and deliberating, or between these and the search for more information, is also a choice. It should therefore be a matter for decision theory to answer. The answer, put mathematically, is so involved that it has been worked out only for the very simplest cases. Nevertheless, this was a victory, and so, if decision theory can be incorporated into the neo-Baconian picture, then the latter is thereby greatly improved.

Usually, the argument that decision theory is acceptable to the neo-Baconians is based on the view that game theory is kosher by their book, and on the strength of the erroneous theorem that the two are the same. (The two coincide under some stringent conditions.) Game theory is an idealization of poker: there are stakes, situations, probabilities, and choice, and at least one other human: the opposite player. Of course, this is, in principle, nothing new: we had sets of states, preferences, etc., and a part of any given state of the world is, of course, other people. Alternatively, a game can be played against Mother Nature, who says yes or no to my choice of prediction, namely who complies with it or not. Of course, this invites protest: Mother Nature has no intents by definition. (Mother Nature is the personification of naturalism, which is the depersonification of Nature, we may remember.) But let us not debate details: this criticism is minor – if Mother Nature behaves as if she was playing with Rational Man the game described in Games Theory, then the neo-Baconians will be utterly delighted. But, it turns out, only a very small part of game theory has been worked out, most of it concerning zero-sum games, i.e., games where the total gain and loss is zero, as in the case of poker played between friends (in a casino the sum is not zero but a negative number, unless the house enters as a participant in the game). Nature may play games, but if She were rational and if She lost every time we gained, She should have stopped playing with us long ago. Of course, this objection in answerable too. It is mentioned here as a token warning: the neo-Baconians claim to have presented a universal calculus, and perhaps they have; yet the more one studies the calculus the more one sees how seldom it is applicable. It is more a matter of principle than an idealization. Here we have arrived at the first criticism that is deadly: neo-Baconians do not present an ideal case as idealization; they have not a simplification of reality but a utopia – not an idealization but an ideal.

2. IDEALIZATIONS VERSUS IDEALS

To notice the difference between idealization and ideal quite clearly, let us consider the game of tic-tac-toe or naughts-and-crosses. It is a game with a few moves, since each side puts in turn his sign on one out of nine squares obtained from a square divided to three by three, and the one wins who puts a row, column, or diagonal. This, however, does not matter much as the set of many moves may be viewed as one move. Children play the game until they discover with little dexterity the way by which the one who makes the first move can doubtlessly win. They then stop playing the game. In principle chess is no different, but the principle is inapplicable, not even with the aid of computers. In both games, then, games theory is no use. So let us take poker. This leads us to an already mentioned psychological criticism, based on the observed fact that people do not compute their probabilities properly. This may be a slight deviation from the ideal, and so of little force. Do poker players who make the effort at playing well calculate probabilities? Yes, though very few do so with the highest available precision. Do they use games theory? No. Let us take consumers, then. Do they order their preferences according to the theory of consumers' preference? No. Do economists worry about this? No. Why? Because they only study aggregate consumption. Why, then, does the theory discuss individual consumers rather than aggregates? This question is crucial. Economists do not discuss it but suggest it has to do with the scientific foundation of their work. When they want to make their theory of preference even more scientifically founded than it already is, they talk of revealed preferences instead of mere preferences. This, of course, is a joke in poor taste, since not all visible behavior but only actions reveal preferences; the difference between voluntary and involuntary behavior is exactly in the ability to deliberate and choose between options, not the behavior that is possibly the final result of deliberation and choice and possibly mere automatic movement. Once we get away from the observed facts into the realm of theory we can ask how many alternatives does Rational Man have, how many of those he can safely ignore without ceasing to be rational, and how much time for deliberation he has, and how much time he has to deliberate about this, and so on.

Books devoted to the subject vacillate. At times they present Rational Man as a distant ideal, based on the ideal supposition that Rational Man has unlimited computer time at his disposal, that he can calculate how other Rational Men's decisions will affect his set of options and vice versa, and more.

This would be a silly joke but for two arguments, each of which belongs to the most valuable treasure of our cultural heritage. One of these two arguments is Plato's, and concerns the place of ideals in the theory of rational action in general: in order to act rationally we have to imagine the ideal we strive for even when we know – as we all do – that ideals are unattainable. For example, when we want to act in the sphere of politics, it is more open to us to act rationally the more we know what we want. The ideal state, then, as the end of political action, is what makes that action rational. And, of course, neo-Baconians do not agree with Plato's ideal state. Rather, they want the ideal Rational Man, and whatever political system is rationally ideal can only be found if we know what is the ideal of rationality.

And here comes the second of these arguments. The Baconian ideal was at first but an ideal research scientist. Bacon himself notices (*Essays*) that the radicalism he recommended for the study of nature, once applied to society – to the Law, as he called it – must be subversive to the state; and so he recommended against it, since politically he was very much a conservative. But the great Baconian John Locke thought otherwise. Locke was taken as the greatest authority on these matters, since he was the leading Baconian and the closest friend of the greatest scientist of all times, Sir Isaac Newton. And he was a political liberal. And so, he tried to present an ideal political system suitable for Rational Man. Many great philosophers in the eighteenth century viewed themselves as commentators on Locke's philosophy, among them Bishop George Berkeley, Voltaire, and David Hume. Hume's friend, Adam Smith, presented an economic theory suitable for the society of Rational Men: their government only plays the role of the policeman or nightwatchman: since each of them acts in order to achieve his own goals as best he can, all that the government has to do is to prevent crime. The question, we remember, how best can one act in the light of one's circumstances, is complicated by the fact that every one person's circumstances include other persons and their aims, and this makes the picture most complex. As Adam Smith has argued, however, there is a ready-made computer that facilitates these computations greatly: the market mechanism. To use computer language, the market is an analog computer whose imput are products and money, and its output are budgetlines and prices. The chief problem of the economy as a whole is a problem of the aggregate consumption; it is, How best should we re-allocate resources so that more consumers will benefit? The answer is given by individual people's choice of what to purchase at what price. For, when they pay more for a product, they give incentives to individual producers to produce more of it, etc. Hence the best we can do to solve the problem is to leave things alone: the economy needs no more than individual rational choices: the most rational conduct of each individual makes the whole economy most rational as well!

There is no doubt that this theory is ingenious, that liberal modern economic theory is but a ramification of Adam Smith's Baconian or

Lockean economic theory, and that the theory – often known as micro-economics, or as the general equilibrium competitive model, or as the neo-classical model, or the Chicago school's model – is breathtaking in its beauty and sophistication. There is also no doubt that it is not an economic theory but a political theory which says, let the government see to it that the economy operates utterly freely – with no political constraints. Presenting the neo-classical theory as if it were an economic theory, or as micro-economics, makes it a political weapon in a situation that is far from the ideal it depicts or recommends and hence a questionable weapon, to put it charitably, whose use may be meretricious. The question hinges on how much it matters that the theory describes an ideal which does not exist.

Economists are quick to notice that the neo-classical model assumes the state to be a mere policeman or night watchman, whereas in our present modern industrial societies the state is a major economic factor. It might be claimed that this does not matter over much if the state behaves like a rational individual – like a producer or a consumer – in the competitive open market. And one might demand that the state should be first minimized and then forced to act competitively. Both these demands are impossible to meet as long as the mint is a state monopoly. Some philosophers write long tracts in an attempt to overlook this criticism in the hope that it should go away. This is not very rational. David Hume, Smith's best friend, argued that money does not matter since not prices but relative prices control the market, so that printing more or printing less money has no influence on the allocation of resources. It follows that inflation or deflation do not affect patterns of production and consumption. This conclusion is empirically refuted to everybody's satisfaction. Hume's argument has been improved in many different ways, all known under the collective label of the quantity theory of money, which label covers a multitude of interesting ideas as well as of errors and of confusions that cannot be disentangled and discussed here. The issue, then, is still open.

Neo-classical economists debate the question, from time to time, how much they need concede to the critics who show that he who prints money is no mere watchman. But many sorts of governmental functions make it a force different from both a watchman and a mere force in the market. The one most obvious in this context is the government as a consumer of weapons. For, clearly, this consumption depends on the political configuration, whereas the theory in question suggests that the political situation is exclusively determined by the rational behavior of individuals. This, however, calls for a very complex debate. We will therefore leave this function of the government. There are, after all, many others. Let us notice only three which happen to be of the utmost importance, innovations, education, and migration.

Much has been said about the fact that Adam Smith overlooked innovations. Clearly, his intention was to the contrary: the market, he felt, is the best incentive mechanism for the implementation of innovations. He did not claim, one should notice, that the market offers incentives for discoveries, only for implementations. On the contrary, he felt that discoveries are products of idle minds, so that the market should not be so very efficient as to make everyone work. Still, he had no trouble assuming that the fun of discovery is its own reward, but the wages of implementation are best decided in the open market.

It is a fact that some innovations are implemented in manners sufficiently similar to the way Adam Smith has envisaged them so as to consider them unproblematic for his theory. Any innovation that is sufficiently distinct and which can be (patented and) implemented at a sufficiently low initial investment may at its earliest stages follow his ideas. But, as everyone knows, patenting is not a Smithian idea, and even a patented innovation can often be implemented only by big concerns, not by small firms. This is so either because of the high initial investment that implementation calls for, or because a small firm cannot market its products without the use of a distributor and distribution is the monopoly of the big concerns in many established industries in developed countries, or simply because a small firm cannot easily guard its patents.

It is an irony that the ideal of Rational Man has no room in it for innovations, scientifc, technological, or artistic! And for no lesser reason than that it is an image of the ideal rational individual who, as ideal, equals his neighbors sufficiently to preclude a discussion of innovations that are peculiar to the one single individual who originates one single unique innovation. (Innovations do not aggregate.) The ownership of ideas cannot be postulated in the Baconian ideal, and even if it were, Baconianism takes facts to be unproblematic and so can hardly offer a criterion for novelty, a criterion, that is, to decide which idea, if workable, may be patented, and which not.

The anti-Baconian fideist philosopher Michael Polanyi has written a still not sufficiently known yet very significant short essay about technological innovations. He said that patent testers are officials whose job is to decide what useful fact is a novelty and thus patentable, and what is not. He did not deny that patent officers may be in error; he denied that there is – or can be – any criterion for novelty (though novelty is a criterion for patentability, of course). He insisted that the patent-tester's intuition, his hardly formulated knowledge, his personal knowledge, is the final arbiter of such matters.The Baconians do not take up Polanyi's challenge: they cannot do so without admitting that facts may be problematic, and they will not make this admission since it threaatens the very basis of their Baconianism.

The status of facts for Baconians, old and new, is the status of the unproblematic basis on which to base all solutions to all problems. This blocks any rational approach to technological innovation: it boasts ability to offer criteria for the employment of given techniques, not criteria for incentives for innovation. Even Sir Francis Bacon was not blind to the desirability of incentives; he recommended the institution of the custom, by now most widespread, of honoring great benefactors of mankind by erecting statues in their images made of gold, silver, cheaper metals, stone, or wood, depending on the significance of their contributions. But how is this to be judged?

Until now it is customary to divide a new technological idea into two parts, the physical and engineering side and the social factors involved which should facilitate implementation. This splitting makes little sense, since, clearly, a gadget is geared to the educational level of its prospective user, his habits and tastes, and these are as much socially determined as the patent laws and the market conditions. The inventor has to know not only engineering but also market factors, patent laws, and so on. The separation of the physical from the social side is, however, easily explicable: the details seemingly in accord with neo-Baconianism are put on one side as satisfactory, and the others are pushed aside to be considered in practice when inescapable, and also to be the subject of researches which ambitious neo-Baconians may conduct in the wish to improve theory. To repeat, there is here an obstacle that cannot be overcome without deviating from neo-Baconianism: one has to admit that some facts are problematic, that some people are cleverer than others, that incentives other than those the free market offers may be desired, that patent laws are problematic and so have no place in neo-Baconian philosophy nor in classical political economy. In brief, technology does not accord with neo-Baconianism. The reason is simple: in an ideal situation there is no room for improvement. We saw this with respect to technological innovation; we can see this more clearly with respect to education.

3. THE EDUCATION OF RATIONAL MAN

The second critique of general neo-Baconianism or its version as a neo-classical economics is that in either garb it precludes any theory of education. For, neo-Baconians assume as their first and primary axiom that everyone is utterly rational. Rational people are, perhaps, already educated. Alternatively, they know how to decide rationally, quite in general, and so they also know how to decide rationally whether to seek education and which education to seek and how much and how often. A rational young professional, for a simple example, has the choice between terminating his education in order to enter the market full-time, splitting his time between education and work in many ways (alternating the two, working part-time, taking evening and summer courses), or postponing entry into the market in preference for continued full-time education. Decisions of this sort are, in principle, like any other decision, open to examination by students of decision theory or of inductive policy, or whatever else one calls neo-Baconianism these days.

Education is problematic to all philosophers except traditionalists, and for the obvious reason that educators are determined to transmit what they know is best in the tradition and expect little or no appreciation from their charges. Baconianism, we remember, is radicalist: it takes it for granted that tradition has hardly anything worthwhile to offer and that traditional education is therefore nothing short of a disaster. Both Bacon and Descartes, graduates of two leading and renowned universities, had nothing but contempt for their education; Spinoza contemptuously refused a professorship in another leading and renowned university (Heidelberg). The greatest and the most famous Baconian, Immanuel Kant (his philosophy is a synthesis between Bacon's inductivism and Descartes' apriorism, and the motto to his greatest work is taken from Bacon's preface to his collected works) was a famous professor who felt that all education is problematic. In his most programmatic "What is Enlightenment?" he said that enlightenment is self-education and self-determination; and in his lectures on education he noticed – what already Rousseau had noticed – that compulsory education is illiberal and so unenlightened and so forbidden. Rather than ask how can education be without compulsion, he asked how can we justify compulsory education, since he could not even imagine education that is free, and this despite Rousseau's detailed description. His justification of compulsory education has brutal logic of the kind that evokes both admiration and revulsion: children should be educated (by force) and humans should not be educated (by force); hence children are not human; they are not human, he explained, since they are unable to exercise the self-determination which makes humans human; and they cannot exercise self-determination, he further explained, since they are unable to exercise their wills, and this for want of discipline. The aim of education, then, is not to transmit knowledge: the acquisition of knowledge is the life-long process of self-education. Rather, it is the acquisition of discipline; and the moment the pupil has gained discipline it is forbidden to teach him and it is then a matter for his decision to continue study or not. So much for Kant on education.

What sounds so eerie to modern ears is not so much the claim that a child is not human. After all, when we are engaged in the debate about abortion, quite regularly we fall into a hopeless debate on the question, are fetuses human? Infanticide was practiced in antiquity openly and in the modern world with little cover-up; only recently did the more civilized parts of the modern world get rid of it.

Infanticide is the outcome of the inequality of the sexes since it hit girls more than boys; otherwise abortion and contraception are preferred to it as a matter of course. Anyone condoning or advocating population control – all responsible enlightened people, really – are forced into the hopeless debate, then, when does an organism become human? As long as the debate goes on along such (irrational) lines, the question, are children human?, terrible as it is, cannot take us by surprise: we are used to such terrible questions and even to worse questions such as, are Jews human? (No, said Hegel in a manuscript work when Kant's sun was still in its zenith: they are merely mechanical.) And it is only when we stop responding with a pious horror that we can notice how inane Kant's response is. It is one thing to legislate, for good reasons or bad, that an organism becomes fully human when coming of age, or at this or that stage of its development, and quite another thing to offer a rational theory for a sharp cut-off: if really a child is not a human being on account of his lack of discipline, then surely madmen are not human either, perhaps also people who suffer from cerebral palsy, Parkinson's disease, or overall paralysis, not to mention drunkards or even people too deeply immersed in prejudices and superstitions to be able to exercise or even possess a will of their own. Kant's answer, like many a logical answer, is more of an indication of the weakness of an intellectual system then of the force of logic.

The problem is still unsolved because it is insoluble. Even the question, how much does the proverbial man in the street conform to the ideal – even this question stands in the way. For, throughout the literature this question is ambivalently answered. Some authors view probability as the natural faculty of every person and ascribe their differences of opinion to different initial (inborn?) probabilities and/or different stock of information (Bruno de Finetti). And, of course, differences of more complex character may also be ascribed to different preferences and to different policies (more prudent, more adventurous, etc.). To be consistent, holders of this view should apply it to babes-in-arms just as well. Do they? How do they handle coercion, then? But then coercion need not be all educational; it can be, and at times is, political. We may oppose coercion in the name of rationality or we may ask how does a Rational Being behave in the face of coercion? And we can do both. Hence, the question of education is still open. Kant's arguments show it is beyond the ken of the neo-Baconian philosophy.

Neo-Baconians cannot abstain from taking a position concerning a vital question just because they have no systematic answer to it. Some neo-Baconians are traditionalists regarding education. This calls for no response. Some neo-Baconians justify the law of compulsory education by the claim that your refusal to send your child to school (notice the paternal authority smuggled in here) may increase your utility, but it reduces mine, since I wish to live in a civilized society. This is an argument that even neo-classical economists give, and quite unhesitatingly and hardly noticing a problem here. One shudders at the thought of justifying coercion by the utility it has for the one who dares use force. But then, again, the neo-classical economists do not mean it. It is but an excuse.

Far be it from any reasonable critic to forbid all excuses: we all make excuses now and then, and with this measure of truth or that, and with this measure of consistency or that. All we need uncompromisingly oppose is the pretense that an excuse made to save a theory is a corollary to that theory.

Even that obvious rule has an exception; it is pragmatism, especially pragmatist fideism, that justifies all excuses, especially concerning important matters. And education surely is important; and most neo-Baconians are pragmatist fideists whose faith in science they justify by the fruits of science. We shall return to pragmatism later on.

The main critique of neo-Baconianism is the Gellnerite claim, we remember, that it is a miserable social and political philosophy. And this critique is based on the claim that neo-Baconianism is pushed towards the social and political philosophy advocated by the neo-classical economists. And the critique of this social and political philosophy involves the claim that governments cannot act as mere policemen or night watchmen as long as they have a monopoly over the printing of money, and the claims that innovations, education, and migration are outside this philosophy. We have seen that innovations call for incentives other than the open-market offers, and that patent laws are both deviations from the market mechanism and inadequate. We have seen that the very idea of compulsory education, being a recommendation for coercion, is opposed to this essentially libertarian philosophy. Some followers of this philosophy today toy with the view of a far-reaching libertarianism in which governments do not print money, and perhaps exert no force other than that of strictly policing, or even a country with no government at all. The question how can such a country defend itself against foreign forces takes us to the field of international relations and to the question, what is the neo-Baconian position in that field? Even when developing the social and political philosophy of extreme libertarianism, however, authors prefer to avoid the question of migration. Once a government of a relatively rich country gives up its control over migration, it will be sure to be flooded by immigrants from poor and from oppressed countries. The theory of market mechanism should allow the market control migration, of course, and, indeed, internal migration in free countries is one of the factors almost entirely left to the free market (except for certain tax exemptions and other governmental incentives to migrate, all of which may be ignored here). Yet had borders been opened, there is little doubt that one could expect serious social upheavals.

Once governments allow only desirable aliens to immigrate, however, another trouble ensues: the process of brain-drain occurs. The people who can best help their homelands by creating and implementing innovations there find it much easier to migrate to rich countries. Also, because technology is taught in a neo-Baconian framework, engineers, whether trained in the modern industrialized world or in poor and technologically backward countries, cannot face the challenge of industrializing societies so different from the industrialized world: they are trained in the social outlook and mores of a society with an extant neo-Baconian ideology, and this means an industrialized country. They cannot see the challenge to innovators that their homelands offer since they can only see their homelands in neo-Baconian terms, and that means that they try to implement in their own country gadgets that are well-known in the industrialized countries, so that the plans they have are both not challenging and frustrating. The situation, in brief, is absurd. But rather than try to understand what has brought them to the absurd, they blame their homelands in a typical Baconian fashion for magic, superstition, and prejudice, and migrate to more modern countries, where their services are better appreciated and exploited. This is no surprise: the education of scientists and technicians is Baconian everywhere.

In other words, the very success of neo-Baconianism in the industrialized world, and more so the fideistic pragmatist endorsement of it as the basic philosophy of science and technology, is also the endorsement of the division of the world into magicians and scientists. It creates a cleavage between the superstitious and the scientific parts of the world and the incentive for brain-drain migration which sustains the status quo.

4. THE INDIVIDUALISM OF RATIONAL MAN

This, then, is the inexorable logic which forces one who accepts neo-Baconianism or inductive logic or radicalism by any other name to endorse the patchwork pastiche that is the social and political philosophy of the libertarian or the neo-classic or the Chicago school. They confuse the idealized, simplified image of the rational man and the ideal society with a minimal state. To straighten matters out, we may ask, can inductive logic be glued to a different social philosophy? The social philosophy described thus far is based on the extreme individualistic method (methodological individualism; psychologism) that postulates no social entities except as aggregate individual characteristics, or as mere conventions, or as a mixture of both. Opposed to this individualism is, traditionally, collectivism. Collectivists think that individual preferences are socially – and educationally – determined, that views are shared through traditions, religions, and national myths. Collectivists insist that social institutions limit individuals' choices and also direct them. Can all these collectivistic paraphernalia be accepted by the inductive logician? No. If the institutions withstand rational examination, they they are rationally justified and so need not be postulated. Otherwise they cannot be justified at all and so the neo-Baconian inductivist will reject them. In particular, why should I – here "I" is Cartesian, the "I" of any rational thinking human – want to be governed? Only so as to safeguard my life and liberty. End of argument.

The above argument is very brief, and in one move. Also, and more importantly, it is valid. Like most valid arguments, however, it does not establish its conclusions but, by drawing questionable conclusions from seemingly self-evident premises, it renders these premises questionable.

Experience suggests that Baconians are incredulous when they hear that someone disagrees with their fundamental Baconian postulates. Expressions like, "if you do not believe in induction, why do you not walk out of a window of a high-rise building?", or "I can't imagine anyone not calculating simple probabilities when crossing the road", are quite common among naive Baconian philosophers as well as among sophisticated ones. One reason for this incredulity is very simple: when a Baconian finds fault with a view he tries to reject it as totally and completely as possible – he hopes to have no trace of it in his system of beliefs. When he hears that someone rejects Baconianism, then, he tries to imagine someone totally rejecting everything Baconians assert, including their claim that people try to act rationally, etc. But on this Baconians are in error: rejecting Baconianism may include the rejection of the idea of a total and complete rejection. And this one has to do. For, no doubt, though false, the Baconian lore has done much for the making of the modern world. Baconianism is, indeed, for better and for worse, the historical myth of the industrialized world, as Ernest Gellner has observed.

Bacon was in error in condemning all alchemists as liars, yet he was right in his disbelief in their claims; he was too harsh on all superstition, yet his condemnation led to the waning of the fashion of seeking a secret in every enigmatic ancient saying, in looking for clues everywhere. Bacon could not start afresh, nor could his disciples. His great follower Descartes demanded that all doubtful assertions be denied – which is obviously absurd. Yet between them Bacon and Descartes raised the level or critical attitude toward tradition and more than anyone else they contributed to the growth of the tradition of self-reliance. Bacon could not put an end to scientific schools and fathered the most dreadful and obviously false myth of science – the myth of unanimity in science. Yet the myth also put a limit to the defensiveness and casuistry which scientific schools were permitted by the scientific public. Bacon's cabbalistic idea that all combinations and comparisons and contrasts will suffice to enable one to find the secret of the universe, provided one begins with no intellectual framework, had a supreme influence on successive generations. This should have virulently clashed with Descartes' demand to establish firmly one – and only one – true intellectual framework, i.e., mechanism, the philosophy that the world is one big clockwork, that the different facts of nature are merely different combinations and recombinations of a set of given cogs and wheels. Yet the Royal Society of London was a follower of both. Bacon's method was that of indiscriminate but clear observations that must be clearly compared, contrasted, combined, and recombined. Descartes had a different theory of method: study each part separately; see that you studied them all; be clear. It looked as if Bacon and Descartes fully agreed. The real difference between the cabbalist tradition and that of the Enlightenment inaugurated by Bacon and Descartes lies just here: the cabbalists were confused and hoped to follow every hint, whereas the mechanists – following Galileo Galilei – were clear and followed only very obvious tracks. The great discovery of the Baconians is the discovery of the obvious. This can become silly, and twentieth century philosophers of the Baconian persuasion foolishly look for the secret of philosophy in the tritest and most unproblematic facts. Little can come out of all this now; yet when in the seventeenth century, rather than think about the philosopher's stone, people started discussing the way billiard balls behave when they collide, something most marvelous was happening.

It is hard to see how impressive the social and political philosophy of the Baconians was, just on account of its great simplicity. Man is rational, rationality gives birth to science, science to new techniques, to technological society, to freedom from want. The economic theories of Adam Smith and David Hume are now covered with heaps of criticisms and corrections. At the time they were the paragons of simplicity. Smith did not deny the historical fact that some economies were based on slave labor, nor did he find it necessary to preach against the evil of man enslaving man; rather, he subsumed slavery under the Baconian heading of prejudice and superstition: it is sheer superstition to think that one can gain more by putting one's neighbor in chains than by letting him act as he will! Slavery then is not so much sinful as stupid!

Adam Smith's most celebrated disciple was Jeremy Bentham, best known to the general public for his utilitarian philosophy and to the experts for his attempt to abolish the classical theory of justice as retribution, and who was thus able to propose ways to reform the penal system. Bentham compared himself to Newton: he discovered the law of human gravity – men seek friends. Hence there is no need to fear one's neighbor. But should we not fear the tramp? Yes. And what then should we do with him? Make him into a neighbor. How? Give him a house with a large-screen color TV. Is this not more expensive than jail? No. Is this not rewarding a potential criminal? Yes; why not? Will this not encourage crime? On the contrary, it will prevent crime and even rehabilitate the criminal and so diminish crime.

Anyone who finds the previous paragraph paradoxical is not well educated. Anyone who thinks the view it expresses is true is not well informed. The view is not true, but much wiser than traditional views which make it appear paradoxical. It must be hard today to feel the full impact of the paradoxicality which traditional views bestow on this Benthamite view. Let us take a concrete example in order to contrast the older and the newer views; let us take self-service department stores, self-service supermarkets, and so on. Self-service offers incentives to theft. This is a fact. Of course, other things offer incentives to theft as well: poverty, unemployment, and so on. Yet a store owner may feel that whereas he cannot be blamed for poverty and unemployment, he may be blamed for offering an incentive to a thief by making his store self-service: indeed he becomes, strictly speaking, an accomplice to theft when he does so. Nevertheless, the fact remains that even though he offers incentives to theft and even though he becomes an accomplice to theft, as a result he may be financially better off, not worse off. Economists calculate with little difficulty the cost of employing salespeople and loss due to theft and conclude that as long as theft is such-and-such a percentage of the turnover (usually 3%) or less, it is cheaper to employ no salespeople and forego the loss due to theft. Here, again, we see the cleavage between technologically advanced societies and backward societies: the advanced ones are richer and so can afford self-service which is more convenient and which enables them to get richer.

This example should suffice to illustrate the thesis here advanced: the Baconian claim to have no ideology and advocate scientific and technical progress as the best cure for man's ills, as the best medicine for the troubles of the age, whatever these happen to be. But Baconianism is an ideology and an attitude to technology that combines machine making and social organization in a specific way that was tolerated in the classical industrializing societies, but not in the industrially backward world of today. It has outlived its usefulness even in the industrialized world, and must give way to a more holistic approach in order to enable us to solve the large-scale problems of the world.

5. CONCLUDING REMARKS

The enormous value and great success of Baconianism should not mislead us to suggest that what we want is but a modification of it. On the contrary, it allows for no modification, since it divides the world rigidly to the utterly scientific and the utterly superstitious, since it begins with an unattainable clear and simple ideal and then applies it to crude complex reality, since it leaves no room for a sociological approach to technology, to education, and to all the large-scale problems which threaten the barest existence of humanity. In particular, leaving rationality an ideal and thus in the hands of the Baconians, imposes irrationalism on all those who cannot abide by Baconianism. Baconianism makes too radical demands on the individual;

it leads to unacceptable old-fashioned social and political theory. What we want is a better theory of rationality that should also lead to a better theory of large-scale problem-solving activity and to related technologies to implement new large-scale solutions.

CHAPTER 6

THE PRAGMATIST MESS OF POTTAGE

As much as Baconianism is traditionally important and seems still to be powerful, it may be noted that traditionally Baconianism opposed pragmatism, whereas today it gains support from pragmatist quarters. Today both rationalism and irrationalism are supported by pragmatist arguments. This invites us to take a closer look at pragmatism. This turns out to be a messy job. For, unlike magic and classical rationalism, pragmatism is more of a mixed bag. To take but one example of how troublesome it is, let us remember the fact that both magic and classical rationalism may be, and often were, endorsed on pragmatic grounds. Indeed, pragmatism is so very vague that even presenting it is very problematic. The starting point may, perhaps, be the failure of all other philosophies and the recognition, in despair, that life goes on even after the death of philosophy. Alternatively, pragmatism may have a measure of toleration and commonsense which the alternatives to it are so dangerously devoid of. This last point seems to be the forte of pragmatism as well as its historical origin. It may be advisable, then, to start from the problem of toleration: why is it so difficult to be tolerant and why is it so desirable?

1. THE THEORY AND PRACTICE OF TOLERATION

It is not clear that toleration was ever taken for granted by any culture, or that intolerance is the rule, or what is the basis of tolerance or of intolerance. Let us remember that even the concept of tolerance or toleration is not very clear. Strictly speaking, when we speak of tolerance we speak of the requirement to permit our neighbors to practice any religion they want. Intolerance was practiced in Renaissance Spain, where heretics of all sorts, whether Jews or deviants from the party line, were burned at the stake for no other reason than that they did not conform religiously.

Even this so very strict definition of tolerance as the minimal religious tolerance that so easily evokes sympathy with tolerance and horror at intolerance – even this definition as worded above is problematic. Many parents refuse their children medication on religious grounds and so, to our technologically minded public opinion, are culpable of criminal neglect and of the obstruction of life-saving services. It is a brute fact that certain Christian sects oppose all medication. Some social philosophers and sociologists oppose technological development, or at least the application of mathematics to social affairs: they claim that overmechanization and mathematical methods in social science endanger the uniqueness of man, the divine spark in man, etc. And at least one leading contemporary philosopher, Mario Bunge, accuses them of the same sort of criminal neglect since, clearly, the application of mathematical methods to social affairs may save lives no less than medication. (Think of the problems that ambulance services may face.)

We have here, in this example, a glimpse at the new approach to technology: whereas the Baconians had hoped that the discovery of new compounds that might combat deadly diseases will lead humanity to victories in public health, it is public knowledge today that this is false, that often we can bring medication to the docks of a city in need for it, or even relief to a starving nation, and be unable to deliver and administer the medication or relief food where it can be of vital service. This is due to political factors – at times due to a whole way of life, at times due to a religion; and, in any case, attempts to overcome the obstacles in question are often blocked in the name of toleration.

We can try the following alternative to, or modification of, the previous definition: toleration is of any religious belief or of any other belief; in brief, toleration is of any conviction, though not necessarily of any action. This definition precludes even the toleration of religious education; hence it permits a government, like that of the Soviet Union, to be tolerant of religion in her very attempt to eradicate it by repressing religious education. This definition, then, evidently will not do.

What practices should be tolerated? We may say, those permitted by law. Correct, but useless: our question remains, what practices should the law tolerate? We must realize that miseducation is not tolerated by any government or legal system that we know, even though laws, governments, and societies differ as to the definition of miseducation. Thus, the tolerant countries will tolerate certain educational practices that in most countries on earth are strictly forbidden or illegal. Two examples are very obvious: the teaching of comparative religion, and more so of sex education, so-called, namely, a little of elementary biology, sociology, and psychology that pertain to human reproduction; they are legal in tolerant regimes and are positively recommended in tolerant societies. Usually, highly industrialized societies are more tolerant than less industrialized ones, at least on these two matters.

—

The reason for this is not what one would like to believe and what is ever so often claimed. Liberalism, let us now admit for a moment with no discussion, includes toleration. And it is often claimed that industrialization leads to liberalization and/or vice versa or that liberalism is a necessary condition for industrialization or a necessary consequence of it, or something like that. This is not true. Rather, industrial societies are often liberal and tolerant because, historically, industrialization is the result – at least partly – of Baconianism, and Baconians are intolerant of all religions and all taboos, especially sex taboos, especially the taboo on information, whether on religious matters or sexual matters or any other. It is admittedly a fact that (prior to the discoveries and opinions of the Baconian Charles Darwin) Baconians were as often religious Christians as not, and that (prior to the discoveries and opinions of the Baconian and Darwinian Sigmund Freud) they often pretended that sex did not exist. Nevertheless, toleration of religion was for Baconians a two-edged sword. Baconians always defended breakers of taboos, since they viewed the taboos as mere superstitions; they often also defended religious heretics for similar reasons. All Baconians agreed that members of different religious denomininations should listen to each other and hold open discussion, just as holders of different scientific opinions should. Yet whereas all Baconians felt that science will benefit from open debates, very few believed the same about religion. There are exceptions, like the arch-Baconian Dr. Joseph Priestley, known today as the discoverer of oxygen. He wrote a book inviting the Jews to an amicable debate on matters religious. Yet, though he was a minister, many Christians in his day did not think he was a Christian, or even religious at all. (He was a Unitarian minister.)

The above discussion illustrates that we can offer examples of degrees of toleration and discuss the views behind them without having (as yet) an adequate definition of tolerance. There is a simple reason for this absence: we – most educated twentieth century people, that is – want to tolerate as much as possible, but not blunt immorality; yet what is blunt immorality much depends on our views. The introduction of our views makes the accepted idea debatable and permits as possibly relevant all sorts of theories, so that we cannot agree about the extent of our disagreement about what counts as that blunt immorality that need not be tolerated.

Baconians should object. (Whether they do is another matter.) They should say that science can offer a framework that is absolute, since science is the one and only framework to be endorsed by all rational individuals: science recognizes no boundaries of religion, nationality, or party. It can limit the debate on blunt immorality, then, and help bring it to a happy conclusion, they think.

This is how Baconianism, even liberal Baconianism, turns out to be less tolerant, not more tolerant, than other modern philosophies. And this is how Baconians, serious and sincere Baconians, became worried about scientific intolerance. For example, consider John Stuart Mill and (his godson) Bertrand Russell. They tried hard to reconcile their toleration and their scientism. Other Baconians, who also found their own Baconianism an intolerant trap, turned pragmatists. The official father of pragmatism is the arch-Baconian American philosopher, William James. Bertrand Russell was most ambivalent about him, showing toward him at times great respect and at times frank contempt.

Consider the Baconian liberal philosophy. Its liberalism constitutes the admission of any individual's preferences without any debate and without any need to justify it. The Marquis de Sade refuted this idea quite satisfactorily and therefore considered himself a great philosopher. His argument has a name among economists: the dog-in-the-manger argument, it is called. The argument says that the Baconian or Smithian or Benthamite view does not yield the utopian conclusions claimed for it since all too often one man's meat is another man's poison, and not by sheer accident: people's preferences in principle cannot be harmonized because some people wish to harm others, some wish to be superior to others, etc.

There is an answer to Sade's argument, to be sure: every argument is answerable, though the answer is not always very convincing. The answer neo-Baconians offer to Sade is from human nature: sadism is quite unnatural. Perhaps. What does this amount to in political terms? Either that neo-Baconianism will be applicable only when the preferences of (almost) all will be natural enough, and thus fit the classical theory well enough. This shows that Baconianism is intolerant, though possibly quite liberal. Or that we need laws to enforce Baconianism, by excluding preferences unnatural according to the neo-Baconian view – perhaps by re-education. This is a switch: from government as a night watchman to compulsory re-education for the Baconian theory of rationality (*Clockwork Orange*). This shows that Baconianism is neither tolerant nor liberal. Many neo-Baconians gave up the argument from human nature, the theory of the natural goodness of Man, as Joseph Schumpeter, the leading historian of economic thought, has called it. This risks the whole edifice. Bacon has taught that human reason is naturally right and so its wrongs are due to superstition that is corrupting. And since reason is both naturally good and Man's most important and differentiating characteristic, Man is naturally good. This is Locke's conclusion. If it is to be rejected, it seems that the principles from which it follows must also be rejected. How else can Baconians respond to Sade's argument? F. A. von Hayek, for example, Nobel laureate in economics and grand old man of the neo-classical or Chicago school, has rejected Bacon's theory of rationality in favor of Popper's. He still insists that the market mechanism is the most reasonable social regulatory mechanism, but he admits it is not perfect and so in need for legislation to protect it. Hayek thinks that the market mechanism is the best instrument to inculcate liberal values, and that legislation should only supplement and complement it. This is obviously evasive. For, clearly, legislation gives the state functions other than the night watchman's, and so the question is not whether the state may intervene in the economy, as the neo-classical theoreticians pose it (and answer in the negative), but rather how much. Both Popper and Gellner take this as self-evident, and they seem to be quite right.

And so, the neo-Baconian alleged liberalism is lost. The allegation is based on the liberal claim that everyone's preferences are given and unquestionable, and that all that the government should do is defend them and force people to keep their own promises and honor their own contracts. But this claim, as Bentham has argued, is based on the axiom that we are both benevolent and rational. It is absurd that the state should be called to impose benevolence and rationality. Clearly, the proper liberal philosophy (offered by Baconian thinkers, incidentally) is David Hume's theory of intellectual pluralism and competition between ideas, and John Stuart Mill's theory of the legal defense of minorities against oppression by majorities. Yet, above all, Voltaire's virulent and uncompromising attack on religious intolerance has played an enormous role in the rise of the magnificent modern liberal tolerant society, as was the tradition leading to it and following it, from Baruch Spinoza to Bertrand Russell, and this tradition certainly is Baconian: it takes little insight to notice that the religious oppressor is less rational than the religious deviant, and this even tallies with Baconianism. Clearly, then, Baconian tradition is in part tolerant and in part not. This calls for a remedy. The remedy proposed was pragmatism.

2. THE PHILOSOPHICAL FOUNDATIONS OF TOLERANCE

Pragmatism has priority over tolerance. The first modern arguments for toleration were unsophisticated and pragmatist. The religious wars in Europe led people to declare that it is useless to kill people in order to make them go to the right church; quite generally, toleration was sustained since antiquity on simplistic pragmatic grounds. It is probably a Baconian theory that all primitive societies are intolerant (even though the early Baconian thought savages were noble in their alleged freedom from religious superstitions). Nevertheless it is a fact that many cultures include the dogma of their own obvious superiority over all other cultures; this holds also for some of the most advanced cultures, such as the ones to be found in China, India, Babylon, and Egypt, just as much as in Judaism, Islam, and Christendom. In Greece, we know, this doctrine was positively opposed on the ground that true human nature is the same everywhere, whereas all that is true by mere convention is false – local religion in particular. Nevertheless, many cultures were tolerant, despite their claims for superiority – for example, traditional (not contemporary Israeli) Judaism: it is a dogma of Judaism that non-Jews have fewer obligations to fulfill than Jews, and so their not conforming to the high code of Judaism, though it is their loss, is not reprehensible in any way. Leaving aside such rare cases of toleration, most known cases of toleration are based on nothing more than pragmatic arguments validated by the pragmatist recognition of the relative validity of different intellectual frameworks, often expressed naively as the claim that everyone should worship one's own gods or that everyone should worship God in one's own tradition's way.

The sophisticated form of this very view is usually based on skeptical fideism. And skeptical fideism may preclude some frameworks as not viable – at least it has done so since the days of the advent of Darwinism – but it considers legitimate all viable ones. A more sophisticated variant will view viability as a matter of degree. This is done often in the name of the idea of progress in a Darwinian guise: the struggle for survival selects the fittest, and a new contestant has to struggle harder. This Darwinian version often goes by the name of the American philosopher John Dewey, who is one of the most influential philosophers of all times, since he managed to reform the educational system of his vast country. Yet, clearly, the Darwinian garb is sham; the contestant species or nation or religion which is not the fittest can stay all the same – it will have to disappear only if it competes with a fitter species, nation, or religion for the same eco-niche. For, obviously, a herbivore does not compete with a carnivore, and a hunter does not compete with a scavenger. Hence, ordering species or nations or religions or any other sort of systems in measure of fitness is simply contrary to Darwinism.

Yet for modern scientifically minded pragmatists this very pseudo-

Darwinian idea is essential, since they wish to replace absolute truth with relative truth, and they feel that this may be excessively liberal – to tolerate even cannibalism – unless we have degrees of truth to check the relativity of truth which they postulate.

We have to enter now that minefield that is common to all theories of truth. Experience indicates that efforts to circumvent it put one on a path that sooner or later leads straight into its very center. So let us better get it over and done with. What, then, is truth?

Let us wait a bit with the answer and ask, instead, why is this question pressing? Why can we not circumvent this issue? Can we not make do with the commonsense concept of truth? How does it get dragged in? And why do pragmatists insist on their theory of relative truth as the core of their philosophy?

It is hard to say. Quite a few people, some of them philosophers, some scientists, some historians, have found no need to go into the minefield of the theory of truth. They said that the pragmatic spirit is universal, yet the scientific spirit is not, and the scientific spirit is the restless one that will be dissatisfied with anything short of the absolute truth. If this were so, then, clearly, pragmatists should admit defeat and recommend, in the name of practical success, the search for the perfect theoretical system. This idea, it seems, was first presented by Bacon in his grand attack on the pragmatism of his day.

Yet this idea is false, and has been turned around by the greatest pragmatist of all time, Pierre Duhem. All intellectual traditions tend to claim the status of finality and authority to their axioms, he observes derisively. Nor was Duhem, himself an orthodox Roman Catholic, really opposed to the diversity of views, inside science or out of it. Yet science enjoys unanimity in any given time on any given question, he said. This must keep metaphysics, with the divergence between its traditions, separate from science. This is only possible, he added, if one maintains a separation between scientific truth and metaphysical or religious truth. And in science one must permit change despite the unanimity which scientific theories enjoy at any given time. Scientific truth, then, is pragmatic, relative, and changing. By contrast, the truth which a traditional view claims to itself is metaphysical, absolute.

To date, the only good arguments for toleration are either practical or skeptical. Now a practical argument need not be convincing: it is possibly convincing if it does not conflict with a stronger pragmatic consideration or with a principle. Pragmatism omits any principle other than that practicality is all that matters. If the toleration of even mere opinions turns out to be harmful, then one can say that either suppressing these opinions is even more harmful than tolerating them, or that the opinions of those who recommend suppression may be more questionable than the views they recommend the suppression of. In particular, taking toleration itself as a principle is impossible, since we can hardly formulate it properly without reference to knowledge.

It was Bertrand Russell himself who argued, in his classic *Skeptical Essays*, that we must be sure that we are right before we can impose our views on others, particularly in education. Now, how can we ever be sure that we are right? Can science prove its doctrines beyond any doubt? Is science itself not based on some presuppositions? Indeed, Russell himself offers a workable criterion: accept only what experts all agree about and doubt whatever is disputed among them.

This is not different from classical toleration. The classical essay on toleration, we remember, was a fable told in Lessing's play *Nathan the Wise* of a father who gave the ring (truth) to one of his three sons (three Western religions), yet he gave the other two exact replicas and did not tell who had the genuine one. Lessing, however, was a mixture of skepticism and Baconian Rationalism. Somehow, they all felt that doubt alone is impossible as a guide for action: action bespeaks conviction. And so, Jews, Christians, Moslems, each to his own belief; but when hostilities rise, doubt should curb them.

There is so much wrong in all this that it is hard to understand the conviction that such discussion usually carries. It is useless to recommend both dogmatism and skepticism, both because they do not mix and because the question, the real and pressing question, What should we tolerate? is thereby left untouched. Surely, no one recommends the toleration of religious atrocities – such as cannibalistic religion and the burning of widows on the funeral pyre for religious purposes. And this puts an end to pragmatism! Moreover, the pragmatists declare scientific truth to be pragmatic and metaphysical truth to be absolute. It thereby may tell us about the way science and religion can tolerate each other, but not how scientists can tolerate dissent within science, nor how Christians may tolerate Jews or Brahmins.

It is here that pragmatists protest, and even get enraged. They will declare the above discussion both muddled and superficial: had we begun with an attempt to define truth first, then we might have not come to the above absurd conclusions. What is truth? What does it mean to claim the status of truth for a scientific contention or for a metaphysical one? What is truth?

3. THE DEFINITION OF TRUTH

We have conflicting intuitions about the truth. We have the intuition that our thinking this or that will not change the facts, that pretending that the facts are other than they are will not help us, except in permitting us to live in pretense, in a fools' paradise, where we may enjoy an unwarranted sense of security until disaster strikes. According to this intuition the truth of an opinion is its accord with the facts, or in the agreement or correspondence between opinion and fact. What exactly this means is hard to express or articulate or define. Here again we have an example, contrary to classical rationalist psychology, of a very popular and well-understood idea that most ordinary people cannot articulate. The opposite intuition is as strong, and in the modern world as significant, and is often lamely expressed in the popular remark that there is my truth and there is your truth and there is the truth in between. Thus, when a world-famous psychologist wrote in recent times a book by the title *Gandhi's Truth*, no one understood it to mean the truth that Gandhi has discovered, but rather, what Gandhi believed to be the truth. But what if his beliefs do not accord with the facts?

This is the preliminary situation. Anyone who denies it can be asked to give his reasons. It would be a very stiff demand from any theory of truth that claims explanatory power, that it should take this preliminary situation as its explicandum, as the facts to be explained by it. There are more facts to explain, to be sure, from other public opinions about truth. For example, there is the fact that all followers of the famous philosopher Hegel, including all orthodox Marxists, which thereby also includes almost all official spokesmen of almost all communist governments on earth on matters philosophical, believe the truth to be self-contradictory. Nor are they alone in this matter, since we may include here a Christian leading theologian, the Church Father Tertulian, who has said, "Credo quia absurdum." ("I believe it because it is absurd" is the official, questionable translation; and the absurd is allegedly the trinity in unity of the deity which worried theologians since they unnecessarily worried about its absurd implication that three equals one. In truth, however, Tertulian spoke of the resurrection in the flesh and seems to have merely acknowledged the belief to be absurd.) Assume the theory of truth to be explanatory of some facts. Now, if the explanation has to be logical, then it must conform to the logical theorem that all contradictions are false. And so, it seems, we cannot state the Hegelian theory of truth as self-contradictory, except by declaring it either a muddle or a metaphor. Also, we shall soon see, any adequate explanation of the facts about truth will have to take account of public opinion about matters related to truth, in particular about knowledge and about faith. It will also have to take account of some facts from logic, whether the logical paradoxes or other matters too; this is an open matter since it is question-begging: logic seems to be biased in favor of the correspondence theory, so called. Lest this sound too authoritative for the start of the debate, we should at once observe that mathematical truth is not the same as logical truth, and initial mathematical intuitions of truth may very well be more relativist and even pragmatist than initial logical intuitions. These are unpleasant facts, perhaps, but wishing that they were different will not change them.

So much about the stock of initial explicanda. The theories available, apart from the Hegelian, or rather apart from the Hegelian idea that some contradictions are true, are the following.

First, there is the correspondence theory which accepts one of the above two intuitions as true and rejects the other as mistaken or as muddled or as a mere play on words. By making correspondence the one and only criterion of truth, the correspondence theory makes all truth absolute, and is therefore also known as the absolutist or the absolute theory of truth, or the theory of absolute truth. The reference to this theory as absolutist is more adequate and clear than the reference to it as the correspondence theory of truth, since no one in his senses will deny that certain simple and specific statements of fact are true if and only if the facts are as these statements say they are. And so, correspondence is an element common to all theories of truth; the absolutist theory of truth, however, claims the same for all statements, whether about sunrise in the east or in the west, whether about the geocentric or the heliocentric system of the world, whether about quanta or about the existence of the deity. Thus, an absolutist will have to say (whether he does or not is another matter) that the statement that places the sunrise in the east is true and the one that places it in the west is false; that the geocentric theory is false, and the heliocentric theory is false too, since the sun, far from being in the center of the universe, is not even in the center of our galaxy; and that though both theological statements, the one asserting the deity's existence and the other asserting the deity's nonexistence, are doubtful, one of them is true and one is false, though we have no verdict as to which of the two is true: some say the one is true, others say the other, and still others refrain from judgment from want of evidence or want of interest or both.

Second, there is the coherence theory of truth: any set of statements devoid of contradictions is true by its own light.

The theory thus worded has a legitimate and universally recognized place, and no one claims for it validity outside that place. In other words, the coherence theory of truth that competes with the absolutist theory needs better wording. But let us stay with the theory thus far worded, which will here be labelled as the limited coherence theory of truth. We recognize limited coherence as truth in a Pickwickian sense: we can say what is true of Mr. Pickwick and what not if and only if we are thoroughly familiar with *The Pickwick Papers*, or even with the whole of the Dickensian universe of discourse. Not only is it not true of Pickwick that he was married, since Dickens says of him that he was a bachelor, it is also not true of him that he once gave a lecture on comparative religion or on sex education, even though Dickens never denied that either explicitly or by implication. Anyone writing a sequel to Dicken's classic will have to invent facts about Mr. Pickwick, but they will all have to be true of him, that is to say, in character with Dickens's story.

The limited coherence theory is particularly important in mathematics. A mathematician may develop a geometry without showing interest, qua mathematician, in the structure of the space we live in; when he does show such an interest we say he ceases to be a mathematician and becomes a physicist. Thus, David Hilbert wrote about geometry as a mathematician when he formalized Euclidean geometry and as a physicist when he developed the Hilbert–Einstein equation of general relativity. The division of Hilbert into two persons is, of course, artificial, even a mere metaphor to delineate his interest in one study from his interest in the other. Hilbert was willing and able to change his view of Euclidean geometry, but when he developed the Hilbert–Einstein equation he was not changing his mind about Euclidean geometry, since the truth he granted Euclidean geometry was that of a limited coherence. This, of course, is not true of the history of science at large: Euclidean geometry was traditionally granted the status of absolute truth and this status was recently taken away from it (by Einstein). Similarly, when a mathematician was working on a difficult problem in Newtonian mechanics, let us say the three-body problem, it mattered little whether we would call his work mathematics or physics as far as truth was concerned, but these days we would call such a researcher decidedly a mathematician when we want to grant his axioms truth by the criterion of limited coherence, which we may want to do in the light of the fact that Einstein has robbed Newtonian mechanics its status of absolute truth once and for all.

The above discussion raises a few difficulties or has a few loose ends. First, now that Newtonian mechanics is granted truth in the sense of limited coherence, the same can be said of the heliocentric system of astronomy, and, come to think of it, even the geocentric theory. Indeed, were this not so, we would not speak of the sunrise at all! Thus, limited coherence is limited to a context. How, then, do we choose a context?

The answer to this question will render the limited coherence theory of truth to a coherence theory of truth proper if and only if the answer will deny the absolutist theory of truth. This is a simple and obvious point; yet, it has been noticed by only one philosopher, namely Mario Bunge. The importance of this point is very obvious and pertains directly to pragmatism: pragmatism as such may, but need not, be objectionable: it depends on its context. When the pragmatist says that in principle all contexts have equal rights to the status of the truth, he has gone too far. But when the status thus claimed is not of the absolute truth but of a limited coherence, then there is no objection to pragmatism, even if it is found not very exciting.

Pragmatists deny the truth of the absolute theory of truth. They thereby, of course, deny the status of absolute truth to all contexts. The Baconian or classical rationalist may object and say that the scientific context and only it is absolutely true. The pragmatist will then justly accuse him of dogmatism and ask him how does he know that Einstein too will not be superseded by a future scientist as Newton once was. It looks, therefore, as if there is no other choice: either we grant science – or my own religion – the status of absolute truth, or we agree that each system has the criterion of truth built into it, so that truth is mere coherence within a system. This last theory is often known as the relative or relativist theory of truth, meaning that truth is relative to one's system's presuppositions. Is there an alternative to the classical rationalist's dogmatic absolutism and the pragmatist's indiscriminate relativism?

4. TRUTH AS IDEAL

This is the central philosophical question of the mid-twentieth century: regarding truth, is there a third alternative? Yes: the ancient, skeptical option – as illustrated by the story of Nathan the Wise about the three rings granted to the three heirs: the truth is one, but we do not know it. This story is too crude: rings there are only three and one of them is true; possible systems of which one will be absolutely true are infinitely many, and perhaps none of them is true. The true ring is before our own eyes, we do not know which of the three it is; and we can say with the highest probability that of all the systems we have before our own eyes, none is the (absolutely) true one.

Pragmatists think it is meaningless to speak of a true complete system of the world that possibly we cannot put our hands on, that perhaps does not exist within human language. They point to the uncontested fact that we do not know whether human language is rich enough to include a comprehensive system of the world that is the truth, the whole truth, and nothing but the truth.

An absolutist will object. Of the two contradicotry statements – say about the deity's existence mentioned above, one asserting and one denying it – one is certainly true, and absolutely so by the canons of logic.

The relativist will claim this to be too naive for acceptability: the concept of a deity is, in fact, context dependent: the same statements mean different things within Hinduism, Judaism, or Spinozism. Hence, the absolute truth can only come whole, but we can never have the whole truth.

Hence the question is still open. The pragmatist theory of truth is irrefutable. It is, we shall see soon, too vague even to be criticized. One can only show that it is different from what the pragmatists want it to be, that it does not solve their problems the way they claim it does. But this is no refutation. The pragmatists want, first, to permit everyone to stick to one's own context while recognizing the legitimacy of other contexts; and second, to permit one to be both a scientist and something else, whether in religious matters or in political matters or in any other matter. If pragmatism could do this, it could also permit one to be a Jew on Sunday, a Christian on Monday, a Moslem on Tuesday, a Hindu on Wednesday, a Buddhist on Thursday, a Taoist on Friday and a Shintoist on Saturday. And to which Heaven, then, will one go when one dies?

The pragmatists are not unaware of this criticism. As often is the case with dogmatists, rather than take the bull by its horns they refine their system and so make things more complicated and render obviously valid and deadly criticism into a point open to debate. Their refinement is in terms of degrees of truth, as mentioned above already. It is here that Mario Bunge's point mentioned above is such a scoring one: degrees of relative truth – that is, relative to the absolute truth – may indeed be defined, and as a part of the absolutist theory; not as part of the relativist theory. Without the concept of absolute truth, degrees of truth are meaningless. This last point was already made in a way by the famous philosopher Count Louis de Broglie, known for his material wave equation and for his creation, with Erwin Schrödinger, of (quantum) wave mechanics; he made this point apropos of Pierre Duhem's variant of pragmatism. Duhem was an orthodox Roman Catholic, and hence felt obliged to uphold Aristotelian metaphysics. He permitted himself the luxury of hoping that the repeated rise of the degree of truth of physics will bring it ever closer to agreement with that metaphysics. In this way, de Broglie noticed, Duhem had crossed the line and passed to the enemy camp: he stopped being a relativist and became an absolutist proper. De Broglie's observation stands, and Bunge's discussion ramifies it by adding to it the necessary technicalities of the matter.

Hence, the really exciting debate is among the skeptics, the fideistic skeptics on one hand – who are pragmatists – and the rationalistic skeptics: both agree that absolute truth is perhaps out of reach forever, yet the fideists give up the very concept of absolute truth and the rationalist skeptics keep it as central to philosophy.

This discussion closes the gap between fideists and rationalists almost entirely. For, the question may be raised, what difference does it make whether we hold to a concept of absolute truth once it is conceded to be so remote and ethereal?

The question is put in a pragmatist language, and so it calls for the following pragmatist answer. If the absolute concept of truth does no harm, then surely its vagueness is no less objectionable than the vagueness of the concept of the Buddha or of the Trinity! The pragmatist may object: it is one thing to believe in the Buddha or in the Trinity, since these are competing systems; the absolute is different! It is a system of systems (a meta-linguistic concept)!

This answer is not open to the pragmatist, however, since it will destroy pragmatism – which is also a system of systems. Whereas Buddhism tolerates other religions and Taoism ignores them, Judaism and Christianity expressly reject them. Of course, this rejection can be relativized to Judaism or to Christianity. But a good Jew like a good Christian will utterly object to this relativization. At least the three Western religions expressly claim for themselves the status of the absolute truth! And this cannot be relativized without a drastic religious reform! Hence pragmatism leads to the toleration of Buddhism, not of Christianity – unless it is a system of systems and as such has a higher authority.

To avoid all authority, classical skeptics have recommended to hold all views in balance: once you seem to tend to favor one view, they have suggested, you'd better listen to the advocates of the opposite view, but not all the way. The modern skeptic does not join the classical skeptic: he does not recommend any suspension of judgment: he will allow everyone to hold one's own view. He will also recommend tentativity rather than dogmatism. What, then, is the difference between the sophisticated modern rationalist skeptic and the sophisticated modern fideist pragmatist skeptic?

The sophisticated modern fideist skeptic says that we must live and act: we cannot wait for the truth to emerge; and yet when we want our opinions to guide our actions we must commit ourselves, we must judge.

This is the pragmatist kernel of fideism. It does not hold water. Repeatedly it was claimed that doubt paralyzes. But faith sometimes paralyzes too. Everything sometimes paralyzes. The fact that some of us are not paralyzed has little to do with what we believe or doubt. Commitment has never been shown more useful than tentativity. Pragmatism is plain silly and is taken seriously only because it is an excuse: were Pierre Duhem not a committed Roman Catholic he might have developed his philosophy differently. This may make a world of difference to religiously committed people all over the world. What is that to people who are not religiously committed?

Rationality is not a matter of belief, not a matter of definition, whether of truth or of rationality, or of toleration. Rationality is the readiness to think and to rethink. God knows what this means exactly. But it is a matter of attitudes. To know exactly what this is before practicing it is suicidal. This is conceded by the very same pragmatists who say that one must eat bread even if one does not know that it is nourishing, much less how. Now, the same way one must use one's reason, even if one does not know, etc. For, as a matter of fact, as Karl Popper, the leading philosopher of science today, stresses repeatedly, we often eat poison, and our use – or misuse – of reason has brought about the new global problems we face that may destroy us and all life on earth. The question is, should we remember that our actions may lead to disaster? The fideist says no, or else we will be paralyzed. He thus ends up recommending the very dogmatism which he had intended to combat. Thus, the more reasonable answer is yes, we must remember that we may cause disasters or else we will act irresponsibly.

In this debate toleration has no place. Toleration, the historically strong point of pragmatism, is by now too problematic for the debate between the pragmatists and the rationalists. It ceases to be the monopoly of pragmatism, as it is shared by those rationalists who make no excessive claims to knowledge and to the knowledge of what constitutes rationality.

Does this not amount to the abdication of reason? It does on the premise that reason must be well defined for us to be able to believe in it. This premise is both false and excessively rationalistic. In debate with pragmatists one is better off noticing that the pragmatists permit themselves all sorts of vague concepts, from the concepts within private religious systems and the concept of a system at large to the concepts of contexts and of degrees of truth, while demanding from opponents a high degree of precision. In particular, pragmatists view the more advanced theory within a given science as more useful than the less advanced theory, so that the advancement of science is the advancement of it as a technological instrument. This view, taken literally, is palpably false, and taken as a metaphor is too vague to be of any use.

This, the pièce de résistance of all pragmatism, is an argument which is accepted on the authority of Ernst Mach and Pierre Duhem and the majority of modern physicists. It is a strange fact that the more mathematical, technical, a science is, in fact or in orientation, the more its practitioners accept this argument, now popularly known as instrumentalism: science aims at prediction, not at truth; Galileo was great allegedly because rather than investigate the causes of gravity he found a formula permitting precise predictions applicable in ballistics, etc. And, of course, prediction is supposed to be the same as usefulness. And this view, too, is so very obviously false that its having currency is only due to its alleged usefulness (for its advocates), not due to its alleged truth. It is obvious that formulas that have only predictive power are looked down at as mere instruments. In science the desire to explain why a formula works overrides the fact that engineers may possibly be satisfied with it. Today, as in the last century, Newtonian mechanics is still found useful – more useful than Einsteinian mechanics – by most engineers and technologists. When engineers do use Einstein, they normally use his earlier and cruder theory, not his most sophisticated theory. We have, in point of obvious fact, theories/of high scientific value and low technological value, and vice versa, and we have scientific ideas developed with an eye for technological application and ones developed in complete disregard of technology, in both cases at times with great scientific or technological value and at times not. Every moderately knowledgeable student of these matters is knowledgeable enough to think up examples for all the many categories just mentioned. Why, then, are sophisticated people so obtuse and crude when they talk about science and technology?

It is the Baconian identification of error with sin, endorsed by most great classical philosophers, including Descartes, Spinoza, and Kant. As Karl Popper has observed, once we have a clear idea of the demarcation of truth from falsity, then the endorsement of falsity is deemed avoidable and so judged condemnable. The idea that science is truth and that there exists a method of generating science was endorsed by these people, and they thereby made condemnable not only scientific error but even scientific stagnation: any researcher who makes no true discovery is thereby at fault. This, of course, he has in common with the magician – any magician who fails to deliver the goods, we remember, thereby proves himself to be a faulty specimen, a person unworthy of holding magic powers. Once we agree that not all errors are sins, we may freely speak of scientific error without thereby launching a condemnation.

5. CONCLUDING REMARKS

The concept of truth has immediate religious and moral overtones. Classical rationalists try to overlook them: they held religion in contempt or viewed it as a second-class citizen in the kingdom of the spirit, and they considered morality an aspect of rationality and rationality as natural. Pragmatists begin with the religious overtones of the concept of truth and cling to them in preference to the factual meaning of the concept. They thus permit us to lie in the name of our clinging to the religion of our fathers. They mask the helplessness that is an aspect of this clinging – and by the claim that faith is practical, that action is impossible without belief, although by faith and belief they mean organized religion, not considered opinion, and organized religion is increasingly one of the two chief organized obstacles to human progress (the other being communist imperialism). That they lie in favor of clinging to their religion is not out of character, however, since they make their religion their criterion of truth. And when the organized religion they deceitfully support is organized science, the mess is intolerable. The appeal to the usefulness of science is of no avail here – since it makes deceit about it quite dangerous.

The popularity of the confusion that pragmatism is, especially among practitioners of advanced science and technology, is a subject for further investigations. Yet we can say already now that classical rationalism and pragmatism are the two poles of the myth of science at the present day. Leading scientists and technologists exhibit this in their table-talk though, we should remember, off-the-cuff conversations are like detectives' clues – they are invalid as evidence but may at times be highly informative as leading to such evidence. The one who listens attentively to the table-talk of scientists and technologists cannot fail to notice their ability to talk in a major key and in a minor key, where the major key suffers from excessive self-confidence expressed as classical rationalism and the minor key suffers from excessive defeatism expressed as self-deception in the form of stock pragmatist arguments. In public these leading lights speak differently from the way they speak in private. It is obvious that addressing large publics, usually on public occasions, regrettably still imposes a discipline very different from that which the workshop does: leading lights, being leaders, fancy themselves as public spokesmen and so as politicians, while un-critically endorsing the view that politicians have poetic license and even a poetic imperative (to lie).

What has been discussed thus far are the popular philosophies of the age, classical rationalism and classical fideist pragmatism, and, in particular, the fideistic pragmatist endorsement of rationalism as a creed, the creed of scientism. In popular expression these are the leading views. In fact they are nothing of the sort. They have been discussed as a necessary preliminary, or rather as a preliminary to the preliminary. The job itself has hardly begun.

CHAPTER 7

PRACTICAL PHILOSOPHY AND ITS IMPLICATIONS
FOR TECHNOLOGY

The philosophy of technology may aim at explaining the facts of technology and it may aim at improving the practice of technology in one way or another. This sounds problematic, since the explanation of the facts of technology should belong to the social sciences, or to the physical sciences, since we naturally wish our explanations to be scientific and demarcate any adequate explanation as within the domain of science. Similarly, the effort to improve our technical performance should belong to some practical field or another, whether engineering or administration or any other. What business, then, does philosophy of technology have in relation to the cardinal aims of explaining and of improving technology? It depends on what role one ascribes to philosophy in general and what attitude to technology one adopts.

It is hard to define philosophy, since the different philosophical schools will naturally disagree on this point first and foremost. Nevertheless, we can say that there is philosophy of life and there is traditional philosophy, and the existence and characteristics of these two are much less contested than any known broad definition of philosophy. It may therefore be quite advisable to begin with these and see what this requires of the philosophy of technology.

1. PHILOSOPHY OF LIFE

A philosophy of life is an attitude to life in general and to major factors we encounter in life, whatever these happen to be. As opposed to traditional philosophy, which most people are unaware of, everybody has a philosophy of life; some simple folks are experts at giving expression to the philosophy widespread among their peers and some simple-minded priests articulate the philosophy they wish simple folks to hold, and most educators wish to transmit their philosophy of life and they are usually schoolteachers but at times streetcorner philosophers, like Socrates, or sergeants and lieutenants in the armed forces and even foremen in the factory; and Shakespeare's fools are mouthpieces of his philosophy of life and a testimony to his educational impulse; at times philosophy of life reaches poetic and moral heights, but is is always unsophisticated, even when expressed by the most sophisticated thinkers. (See Bertrand Russell's "Free Man's Worship".)

This is not to say that the philosophy of life is the same everywhere. On the contrary, the philosophy of life of frightened, defensive, inept people is fundamentally different from that of easy-going, open people (like Bertrand Russell). Some philosophers see in the way of life of a person an exemplification, oft perhaps lame, of his philosophy of life; others see in people's philosophy of life nothing but a superstructure on and an excuse, perhaps lame, for their way of life. Both these views of the causal relation between a way of life and a philosophy of life are amply confirmed by experience, and this obviously means that both are empiricially refuted. A person may develop his philosophy of life in order to justify his way of life. A person may also try to live out some inner convictions: in most cultures most of the folk heroes are described as people who overcome obstacles in order to be what they wished to be. The West stands out here as different: in the west heroes usually attempt to do what they wanted to do. The philosopher of technology, the apostle of modernization, the exporter of democracy, must constantly remember this difference and never underestimate it. It is also possible, and at times, however rare, it happens, that the progress of an individual's life and of his philosophy of life intertwine, and they both grow together. These individuals are the really interesting and challenging people, whose lives grow as their thoughts do. Examples are as different as the anti-Chistian Bertrand Russell and the Christian Mahatma Gandhi (who was by denomination a Hindu and a practicing yogi, but who developed a highly Christian philosophy of life, which cannot be discussed here). They are people whose lives, as I. C. Jarvie puts it, are themselves works of art – the art of living.

The main and central questions of all possible philosophies of life concern the meaning of life and its practical implication: what is the point of life? What are we here for? Traditional religions all claim to tell us the answer to this question and the practical significance of their answers. Western religions are attempts to offer mankind answers that will make positive sense of life. This is done on the supposition that an answer is essential for most people in order for them to live at all reasonably satisfactorily and usefully, and on the supposition that most people are better off taking regular advice from their leaders, which they will do only on the authority of some religious teaching or another. Therefore, the philosophy of life of Western religions is not the dogmas they teach in their Sunday schools, but these very suppositions which justify their teaching. This means that Western religious leaders permit themselves to lie to the gullible for the benefit of the public. This is pragmatism, and so, viewed as technology, western religions are poor as compared with oriental philosophies and religions. As much as they are at all consistent, these latter seem to say that life is meaningless. They, too, offer some practical advice, most of it likewise not effective except for the very select few.

It should be remembered that oriental philosophies are anti-technological only in the sense of technology familiar in the West and only when this is cast in the Baconian rationalist mode. Viewing technology in a broader sense, all oriental philosophies, being practical, being concerned with the art of living, are highly technologically oriented. They are, one can say, highly pragmatical – not at all pragmatist, but pragmatical, inherently concerned with the practical implications of their central supposition and full of practical advice about the way to attain the chief aim in life. For, almost all oriental philosophy says the chief aim is the attainment of peace of mind.

Assuming life to be meaningless, the oriental philosopher sees little point in worldly action: quite logically he sees worldly action as bringing nothing but vexation and frustration. He concludes that the only way to attain peace of mind is to train for meditation as the avoidance of worldly action, to attain worldly inaction that is not borne of frustration or of helplessness but of choice: in full resignation yet in full control.

The techniques of attaining resignation and self-control are central to every oriental philosophy known in the West. This makes oriental philosophy include quite a lot of technology in the broad sense of the word, though definitely not in the narrow traditional Western sense.

The theory of peace of mind through meditation, resignation and self-control is not unique to the orient: it is found everywhere and known as asketicism. The asketic philosophy of life is generally misunderstood and often viewed as masochistic; many masochistic religious practices are viewed as asketic, especially by their advocates. The counsel to punish the flesh by administering to the body all sorts of discomforts, from sack cloth and ashes to self-flagellation, are obvious examples – not to mention the abstinence from sex, wine, etc. The Jesuit practice of contemplating hell before dawn for half an hour on one's knees every morning of one's life is a supreme example of the practice of self-punishment administered with the aim of raising one's self-control. Very widespread, though counterproductive, such practices cause much confusion as to what asketicism is. Asketicism, on the contrary, preaches the ease and elegance of not being subject to the temptations that beset ordinary people and that flagellists try to overcome by punishing the flesh: whereas flagellists punish, asketics kill or control whatever temptations that beset others.

It is easy to dismiss the flagellist as a masochist: sadism and masochism are equivalent in their being quite unnatural in the sense that they mix preferences, in their preferring what they also reject, in their ambiguity or ambivalence. True, Freud has argued that ambivalence is common – so common that we all suffer from it to this or that degree. Yet he, too, took the norm (though not the average) to be the clear-cut division into preferred and rejected situations, and claimed that the norm is attainable (and even that he had attained it – a claim amply refuted in any of his important biographies and even in his own writings). Not so with asketicism. We may wish to wish to attain much, and we may wish to wish to attain little and be content with it. This is a manifest fact. It is a fact that oriental philosophers, as well as Western asketics, recommend that we learn to be satisfied with little; that the Baconian Enlightenment has taught us to treat every individual's wishes as given and, upon his say-so, uncontestable; that the Romantic Hegelian reaction to the Enlightenment taught the very opposite of the asketic philosophy: as Hegel said in his pathetic but unforgettable adage, great men are moved by great passions. Can the philosopher of the Enlightenment who boasts of objective disinterest in people's preferences – can he show indifference to the second-level preference between these three options (asketicism, Enlightenment, Romanticism)? Modern economic theory, the most explicit expression of Baconian liberalism, treats consumers' options as given ("exogenous") but insists on the second level on this first-level indifference. Second-level options are not treated there, and so judgment concerning them is not expressed one way or another. Perhaps for the market's behavior we do not need to decide these matters. Yet, the question is, does the theory of the market apply to asketic society? Yes, as long as the asketic society has a free market responding to supply and demand. (The law of supply and demand, so-called, does not regulate either; it is the claim that supply and demand, taken together, regulate the allocation of resources and the distribution of commodities in a manner leading to equilibrium – unless governments intervene). Indeed, a liberal economist may claim that the services of a teacher in asketicism, as of any teacher, are subject to the market's mechanism (if there is a free market in education). We have seen already that this claim is false: the whole business of education does not fit the picture unless it is assumed that with minimal education an individual naturally grows to be an enlightened person in the light of the

Enlightenment movement's view of enlightenment, namely that rational individuals use reason to satisfy their preferences.

Let us also consider the fact that classical economics fails to recognize national boundaries as rational and fails to handle the matter of innovation or of incentives for innovation. Put together these two failings make it hard to say whether products available only in one continent are also on the market in another or whether importing them constitutes an innovation. Nor can the classical theory handle the impact of innovation on preferences, despite the known fact that often enough the appetite comes with the eating, as the folk saying puts it.

What this amounts to is the fact that we do not know how individual preferences aggregate on a second-level preference, even on the assumption that, given first-level preferences, a free-market economy, and a modern technology, consumers' preferences do aggregate nicely and do follow the laws of supply and demand so as to maintain (dynamic) equilibrium. That is to say, even were liberal economic theory true, it would be useless to expect it to work in the direction of what is called modernization. All efforts at modernization undertaken internationally in the modern world in the first three decades of the post World War II era were in the form of economic aid, based on the suppositions that creating industrial options in poor countries will bring about modernization including the usual social and political paraphernalia of the modern world. This is a Baconian bias of an incredible narrow-mindedness, and it rules diverse agencies placed at technologically key points, such as the World Bank and the Federal Food and Drug Administration of the United States of America, to mention but two very important bodies. The world-famous slogan of the Western world, save energy! is a similar Baconian blindness: it is an attempt to show individuals what is rational for them to do in the hope that they will act rationally and that their actions will aggregate. But one reason such propaganda does not work is that no one wants to be its victim: if I make efforts and save while everyone else does not, then I make a sacrifice and no benefit accrues from it. This is why the hope was that the market mechanism will replace preaching: if the cost of energy increases enough so that many individuals will prefer to save energy, then saving will aggregate. Except that the increase of the cost of energy causes inflation and thus spending rather than saving.

It must be noticed, as Keynes has noticed already, that the folly of modern economics is based on the fact that it is viewed as a science rather than as the philosophy of life that it is. As a philosophy of life it is but one among many options available. And hence, the attitude toward technology implicit in it is but one time in one philosophy of life. Other philosophies of life exist and have different philosophies of technology implicit in them.

This is not relativism: the claim made here is not that every philosophy of life is correct; rather that there are a few competing ones going around, probably all of them erroneous to some degree and certainly not more than one of them correct. Rather, the claim made here is that the preference of Baconianism – that it is impartial and indifferent to all preferences – must be debunked. What we are impartial to or indifferent to depends on our philosophies of life in general. And these are and ought to be subject to critical scrutiny from time to time.

One task of the philosophy of technology, then, is to present different philosophies of life and the views of technology implicit in them, and examine them critically. As it happens, clearly, we must reject as refuted both the oriental philosophy of Sun Yat-sen (now practiced in oil-rich poor countries) which recommends the acceptance of Western technology while sticking to local values otherwise, and the Enlightenment philosophy of Bacon which endorses the individualistic world-view, including the desire to advance science and technology. Both are easily criticizable. They are the two most widespread philosophies of life next to the ones that are frankly anti-technological (in the Western sense of the word). Hence, the search for a philosophy of life with an adequate philosophy of technology has hardly begun.

2. TRADITIONAL PHILOSOPHY

Leaving the philosophy of life for a while, we move to traditional philosophy. It is fashionable these days to show a certain disregard, if not disrespect, towards traditional philosophy. This disregard may be based on the radical Baconian standpoint that puts all its faith and hope in science and sees in all philosophy except the philosophy of science a rude competitor to science and so superstitious. The disregard may also stem from the new irrationalist standpoint that sees in all philosophy a strong and unwelcome rationalistic pro-scientific dangerous bias except for some minor, outmoded, old-fashioned strains of irrationalism. Traditional philosophy does indeed have a strong rationalist pro-scientific bias, and for the rather simple and (fairly arbitrary) reason that all other philosophies that are not rationalistic are usually viewed as theology rather than as philosophy, with the exception of more or less recent irrationalism or post-rationalism or trans-rationalism (Michael Polanyi).

The concern of rationalistic philosophies was always with rational standards or choice of standards – whether standards of adequacy of scientific theories or of moral rules or of political rules or of aesthetic rules, etc. The traditional view was justificationist, to use W. W. Bartley's apt expression: any choice must be rationally justified. This led to a crisis in philosophy, with the – ancient – discovery of the difficulties of justificationism: a justification needs defense by a standard of justification, they say; but then they must admit that the same holds for the standard itself. This leads to circularity or infinite regress: either, ultimately a standard manages to justify itself, or there is an infinity of standards required to do the job; either case is questionable. Those who found the crisis quite unresolvable were the skeptics. Their position was viewed as untenable on account of the thesis that doubt paralyzes. This led to pragmatism. Pragmatism, however, justifies, if at all, much too much.

Sir Karl Popper, the leading anti-justificationist and anti-pragmatist philosopher of the age, used to tell in his courses on scientific method the story about a tribe that considered tigers sacred. Unfortunately, he said, that tribe is extinct, as its members were devoured by tigers. This story is pragmatist in its thrust: some beliefs are outlawed by mother nature; others are permitted and thus legitimized by it. Perhaps so. Yet, clearly, this kind of legitimation is too generous in the short run and holds no promise in the long run. That is to say, not all tribes that worship tigers die out. India is now devoured by cows, not tigers; and Israel by pragmatist ideology. And they both manage to survive, even though at a very high cost. Moreover, there is no assurance that India or Israel or Homo Sapiens is here to stay, and perhaps our chances of survival will increase by slaughtering some sacred cows.

One of these sacred cows is, according to quite a few environmentalists, technology itself: they think that unless we in the Western world learn to go back from the use of the automobile to horseback riding, we are doomed. They deem the automobile more dangerous than either nuclear explosion or the population explosion; perhaps they think we cannot handle these troubles without overcoming the temptation of the automobile first, without having first a change of heart. It is very interesting to notice that whereas pragmatism may justify too much – anything that has managed to survive – there is no argument within pragmatism which pertains to the question of the future survival of what has thus far survived. And the way to handle this question is to discuss rationally the question, which of the two forecasts is more plausible, the one which says that gadgets bring heaven on earth or the one which says they bring hell. This proves that pragmatism is no answer to skepticism. Both Baconianism and pragmatism are excessive and their (pervasive) conjunction is silly.

There is a simple argument to clinch this: almost all wise people, East and West, poor and rich, ancient and modern, all speak of peace of mind as a supreme end. There are exceptions, to be sure; nor can we declare off-hand foolish all those who believe that restless spirits are better off than tranquil ones, especially if one happens to side with neither party. Nevertheless, taking peace of mind to be the highest preference of so many diverse people, skeptics and dogmatics, asketics and self-indulgers, civic-minded and self-centered, we cannot possibly justify all of them: their aim is the same yet their techniques are so varied. Do these various techniques work? Equally well? Pragmatists stress the impossibility of discussing views from one framework in another framework: each framework justifies its own presuppositions and is in its turn justified pragmatically. This is false: all frameworks are suppositions allegedly leading to survival and peace of mind; and peace of mind is almost never achieved.

Can peace of mind be achieved better without any known framework? If so, are they not all to be replaced? If there is a new and better way, or even a contender for such a way, is it not worth investigating? Can even the mere hope to find a better way to peace of mind be overlooked in preference of a commitment to one or another traditional way?

The viewpoint from which this attack is launched is the viewpoint of the restless spirit, of course; instead of peace of mind we may put happiness, or, still better, the good life. We may then endorse the classical Enlightenment philosophy or its competitor – Romanticism. The restless spirit of the Enlightenment was isolated: when Adelbert Chamisso's allegoric *Peter Schlemihl* lost all hope of regaining his shadow, he became a traveller and a naturalist; soon enough Chamisso followed Schlemihl and published a book on plants he gathered in Russia. The restless spirit of the Romantic philosopher Hegel was a small man, a Napoleon, who must step on other people, perhaps on thousands of dead bodies, in order to feel tall. Which of the two philosophies is right? What is the nature of Man and what are we really after? Is it really so rational to be rational, to devote one's life to the study of minute facts of nature instead of attempting to join a Napoleon in order to live the life of great deeds as the Nazi philosopher Martin Heidegger recommended?

Philosophy today centers around these problems. The main divide between philosophers still divides those loyal to science and to the view of science as the paradigm of rationality from those who view science and the scientific point of view as too narrow, who consider both science and technology a matter of much exaggerated significance and who await the true revival of faith.

The easiest way to put down science is by having a pragmatist view of science: science is of little informative intellectual value, though it has much technical information that at times may come in very handy. Treating science as an intellectual framework allegedly makes it grotesque: its eternal atoms begin to split, its reinforced classical edifice collapses before modern advances, and so forth. Of course, as technology classical physics still stands, and engineers still find it very handy, but claims made for it as the rational intellectual framework have been superseded. This is pragmatist anti-science.

The easiest way to prop up science is by insisting that its theories are the best we have and the most rational to accept. After asserting this in contrast to all the unscientific philosophies, a philosopher of science usually pulls up his sleeves and works on the rationality of science that makes it so very credible. Here traditional philosophy returns to the foreground: rationality is justification and justification equals proof; if not proof of finality, at least of highest credibility; if not of highest credibility, at least of highest probability; if not objective probability, at least subjective one; if not even that, then at least expected utility of current predictions. The lifetime of one person can be devoted to a pilgrim's progress along this path. During most of his adult life Rudolf Carnap was the leading philosopher of science. He began with verification, moved to probability, to subjective probability, to utility. Had he lived longer, he might have discovered classical economics. Indeed, though he was a socialist, his magnum opus, his book on the logical foundations of probability published in the mid-century, includes one rule of rationality, and it is addressed to businessmen! Of course, he meant men – and women – of action. He began by defending science, but in that very book he failed and fell back on defending scientific predictions instead. Whereas the pragmatism of those who belittle science is bold, that of the apostles of science is rather pathetic.

Of course, those who belittle science do so in order to belittle scientific rationality. They then either take their own parochial intellectual systems for granted and count as mere theologians or, aspiring to the higher position of philosophers, they try to justify their own systems of belief. Not having scientific rationality at their disposal, however, they can do so either from within, which is of no use, or by a fideist pragmatic argument. This, finally, makes them pathetic too.

Nevertheless, the peculiarity of the philosophy of science should not be overlooked, especially here, since it has immediate corollaries for technology. A non-scientific intellectual framework comprises some religious, national, or political general suppositions (the word "ideology" comes handy here: it is vague enough to cover all three and more and has a welcome halo of fuzziness around it). It is tacitly assumed that people believe what the framework asserts, that they suspend rationality and make a "leap of faith", as Kierkegaard has put it, or a "retreat to commitment" as Bartley has. That is, whether the fideist philosopher believes in God or in the socialist motherland is irrelevant; he literally is a fideist, however; that is to say, he means it literally when he recommends that his reader believe in God or in the socialist motherland. Not so the pragmatist philosopher of science. He, if he is any good, takes pain to explain that the axioms of physics do not mean what they say, that they should not be taken literally. This is, of course, a very sophisticated matter, and quite amazing to boot. Let us take it slowly, then.

3. PRAGMATIST PHILOSOPHY OF SCIENCE

The main interest in emptying science of its theoretical content – and thereby of its purely intellectual value – is better posited before discussing the way this can be done. For, if the end of the exercise is desirable, then we may try and try again until we find the best means to achieve that end, and if not, then no matter how good the means, there is no point using them.

The interest of the fideist was clearly stated by Duhem: wanting his religious commitment to harmonize with science, he preferred science to be intellectually quite uninformative. To his non-religious colleagues, however, he offered two different interests. First, the interest in unanimity: if science only asserts facts, then the commitment of some scientists to one framework, of others to another, and of still others to none, is of no concern to science. Second, the interest in avoiding all error: since theories cannot be justified they are probably false, and so, rather than embarrass science let us banish all theories from its courts.

The question then is, how can this be done? The answer turns out to be very sophisticated. The clue came from traditional philosophy of science, from the view that already Claudius Ptolemy had about his own astronomical theories. He was convinced that nature is simple yet he offered complex theories which, he claimed, were fictitious but very useful. Hence the names "fictionalism" and "instrumentalism" and "as if" and "façon de parler".

(Since our image of technology is so gadget-laden, we may have to remind ourselves that astronomy was terribly important for technology. Astronomy was terribly important for navigation and for practical calendar computations. Now calendars are of religious significance, but only technologically oriented societies pay so much attention to precise calendar projection that even their religious leaders have to attend to this need.) How, then, can science be robbed of its theoretical content? How did Ptolemy do it?

Committed to a metaphysical theory (Aristotle's), Ptolemy declared his own theory fictitious. This is not very sophisticated and leaves open the vexing question, if one's metaphysics is true, how come not it but rather some other theory is successful in its application? This question led Copernicus to consider the state of astronomy as something reaching the status of a scandal. When the position of the Church of Rome was so weak that it had to fight back and establish its authority, one of its worst leaders ever, St. Robert, Cardinal Bellarmino, brought about the burning of Giordano Bruno on the stake in the year 1600 and threatened Galileo a decade or so later, demanding that Galileo not assert the truth of Copernicanism until permitted to do so and promised permission upon receipt of proof. This same Bellarmino, as a Jesuit leader, was at the same time in charge of the introduction of Copernicanism into astronomical technology while keeping it out of science: science requires proof, technology only utility. As he put it, Copernicanism is philosophically false; it is a mere hypothesis – or, in modern language, it is a working hypothesis.

All this is not satisfactory. A working hypothesis usually enters the periphery of a researcher's work. A scientist concentrating on a difficult problem may take for granted without question for the time being any hypothesis which he hopes can help him. Hence, the word "working hypothesis" designates a lack of interest and low critical attention. It is well and good not to try to be interested and critically-minded in every hypothesis at once. But to be not interested and not critically-minded in any hypothesis ever does seem a slight exaggeration.

The situation changed a century after Bellarmino. Science became triumphant and pragmatism was boosted by Bishop Berkeley's criticism and theory of meaning. His criticism of scientific realism showed that, taken literally, Newton's theory is mathematically highly problematic despite its technologically stupendous success. His theory of meaning came to show that strictly speaking concepts gain their meanings from experience, in a manner not obtainable by key concepts in Newtonian mechanics. As a theory proper, then, Newtonian mechanics is meaningless; it does not really

exist in the first place. It is nothing, said Berkeley, but a set of tricks – rules
of thumb that are very useful in practice.

This is still not sufficiently sophisticated. It becomes more so when a fuller
account of Berkeley's pragmatism is taken. For, he was sincerely religious
and believed that nothing is as nearly important as doing God's will. This
simple point makes the study of science and the investment in technology
quite marginal to the real business at hand. Yet, since Berkeley loved science
he granted it a religious function, and in this way. In order to do God's will
we must know it, and He commands us both through the Holy Writ and
through our senses. What we see, then, are messages from God. This is but
another version of the Renaissance theory of the two books of God, the Bible
and the Book of Nature. Yet its use as a theory of meaning and of worship
makes it both more beautiful and more interesting.

The idea that science equals proof was the common view. Realists,
whether Copernican, Baconian, or Cartesian, those who believed that proof
is possible, saw in science a supreme intellectual import; the fictionalists or
instrumentalists were skeptics who offered a fideist escape from the skeptic's impasse and then deprived science of its ability to prove, and hence
from its theoretical import, and hence from its intellectual significance. The
realists were militant because organized religion was militant and because
they soon endorsed the Baconian comprehensive rationalist philosophy that
leaves practically no room for religion within the limits of reason alone, and
no room for views or opinions or beliefs outside these limits. Soon science
became triumphant, with mathematics and physics as evidence of victory.
On the authority of Galileo, Newton, and, finally, Kant, it was taken as an
article of faith that the Book of Nature is written in the language of mathematics: once the language is known, the book can be read with ease.
Mathematical formulation lends scientific theory its scientific certitude,
then; and the aim of each researcher is therefore to discover and to put his
discovery in a mathematical language. (One may easily notice a logical
contradiction here, masked by the word "therefore": if mathematics is the
language, discoveries should appear in that language, not be recast in it.)

What is mathematical language? It is nothing but the application of the
axioms of mathematics, and these are, of course, quite obviously certainly
true. Unfortunately, however, the only axiomatized part of mathematics
available before the advent of the new mathematics that is so revolutionary
as to baffle all philosophers, was Euclidean geometry. And this geometry
was problematic: even Kant spoke of some sort of non-Euclidean geometry.
In his epoch-making Inaugural Address of 1770, he said that non-Euclidean

geometry is conceivable but forever useless as a tool for scientific research. Yet non-Euclidean geometry was developed (as a reaction to Kant; all nineteenth century geometry, says Bertrand Russell in his youthful *Foundations of Geometry* of the turn of the century, was developed in reaction to Kant) and the story evolved with a logic of its own. First Karl Friedrich Gauss suspected and then Felix Klein showed that conceptually a mathematician cannot decide which geometry is true: this requires the judgment of experience.

And why not? Already in the early nineteenth century Gauss proposed the idea of empirically testing geometries – a crucial experiment between them. Yet there is a flaw here: if science must be mathematized and mathematics must be rendered scientific, then the ground slips away under the feet of the anti-skeptical philosopher.

4. PRAGMATISM AND MEANING

David Hilbert was not satisfied with Klein's proof: it does not go far enough. Hilbert tried to show that Euclidean geometry is at least conceptually perfectly in order – at least that it is free of contradictions. At the turn of the century, at the same time that Russell wrote his book on the foundations of geometry that turned out to be the summary and the closer of the era, Hilbert began a new one: he said, let us view Euclidean geometry as a game, as a set of chess pieces where each word permitted is a chess piece, each geometrical theorem a permissible state of the chessboard, and each inference a move from one state of the chessboard to another. And let us see if any permissible move leads from a permissible state to a state that we will recognize as a contradiction. If yes, the system is inconsistent; if demonstrably no, then the end is reached and the system proves consistent.

The idea that axioms may have no meanings of their own was sketched by Berkeley. It was developed by the pragmatist instrumentalist philosopher and mathematician Henri Poincaré. Yet it was Hilbert who offered the first systematic presentation of geometry in such a fashion. This is how meaningless sentences were boldly forced into our intellectual universe. The use of the device proved explosive. Information theory, for example, treats messages as sets of chess pieces devoid of meanings; so does computer science. But let us beware: neither Hilbert nor the others expressly deny meaningless sentences their right to meaning: on the contrary, a meaningful statement, a piece of information, theoretical or empirical, may easily be stripped of its meaning and become a meaningless sentence for a given purpose. Only after Hilbert's technique was developed was the idea reversed: suppose we create a meaningless language akin to a chess game and then wish to endow it with meaning. Can this be done? Will it be interesting?

The bias of the present study goes the opposite way: have an aim, a task, a problem. Then, look for a solution or for tools with which to forge one if none is available. But once this bias is put as a general proposition, it is decidedly false. It is refuted by one of the distinguishing marks of applied science: whereas pure science usually (not always) goes from problems to solutions, in applied science often enough new solutions go around in search of problems! (Example: radio-active tracing.)

This is what happened to Hilbert's game. Once mathematical systems were emptied of meaning, the question of the validity of any one of a set of competing systems disappeared, and it was possible to ask, can we restore meanings while letting the sleeping dogs of the skeptics lie? Yes. We can say that we give meaning to some words in the system by the logic of their situations, without having to worry about any correspondence to facts, and we can give meaning to the rest of the words in the system on the condition that the resulting sentences be true. (Once all words in a sentence have meanings it becomes a statement and so it has a truth value: it is true or false.) In other words, the meaning of words like "point", "line", and "between" is any meaning which makes the axioms of Euclidean geometry true. The axioms, thus, are not statements proper, but statement schemas, matrices. (A matrix – originally a womb – is a pattern with patterned gaps, like a jigsaw puzzle with pieces missing.)

Sentence schemas are known as implicit definitions. Implicit, as opposed to explicit: in an explicit definition there is an equation with a defined word – the definiendum – on the left, and the defining expression – the definiens – on the right, where the definiens does not contain the definiendum. Implicit definitions are intentionally circular: a point is anything which makes Euclid's axioms true, and Euclid's axioms speak of points! Nor are implicit definitions as complete as explicit ones. An explicit definition, properly executed, is in principle redundant: we may always rewrite a sentence which includes a definiendum by replacing the definiendum with the definiens. This is a strict law: a definition not complying with it is – by definition! – not a proper explicit definition. Implicit definitions are different. Not only can the word "point" in Euclidean geometry mean a point; it can also mean an ordered set of two real numbers, or one complex number. And it can mean other things (indeed, all things mappable on the Euclidean plane).

The axioms of mechanics may now be declared meaningless. We may grant them meanings and say that in every case we grant the equations meanings and the results are satisfactory – for example in navigation, or in sending a sputnik, a weather satellite, or a man to the moon – we stick to the meaning granted. Otherwise – for example in exploding nuclear weapons or in planning nuclear plants for peaceful purposes – we deprive the equations of the possible meanings. Also, we play around; we try different meanings; we remain flexible. With the great discoveries of Abraham Robinson mathematics recently endorsed the search for alternative meanings – for non-standard models, to use the logician's terminology. It is now a standard mathematical technique.

The point of this exercise still remains the defeat of the skeptical philosopher: when we play this game we see to it that the equations remain either meaningless or true; never false.

The result is simply vain boasting of no interest and mystification. Suppose we entered an archive, of the type envisaged by a Jorge Luis Borges, or a Saul Steinberg, an archive which contains an endless series of passports. They all follow a simple formula: they contain photos, names, details, and nationalities. Are they genuine or false? We do not know. Even the best border inspectors are fooled, we know. But never mind them. We want to be sure that we, not they, are all free of error. So we stipulate: none of the passports has a claimant, unless a claimant appears who fits all the details in the passport he claims. Then, even if two claimants fit one passport, we say it belongs to each of them, and if a claimant turns out to claim a passport not fitting his description we cancel his claim and deny that the passport in question has any claimant. Of course, in this way we will be free of all error. But the border police will not benefit in the least from this exercise. Moreover, anyone can ask, where do all these passports come from? How come some claimants are successful, some not? Do claimants ever fool us? If not, what means of protection against being fooled do we use? The game we play precludes the ability to even attempt to answer these questions!

The pragmatist philosophy of science, one can see at once, is the insistence on the demand for proof coupled with the recognition of our inability to procure proof. Bellarmino and Descartes declared unproved theory false; Berkeley and Duhem managed to soften the blow and declared it meaningless. Both are fideist skeptics and both view science as a mere technological tool. Their view does not stand the initial requirement: it does not present science as useful, but only admits that it somehow may be useful while denying that it is informative. It also denies that we can ever find out when science is useful, when not, and why. Its emphasis on the possible usefulness of science is not a positive contribution explaining the actual usefulness of science; it is but a perverse way of saying that outside usefulness science has nothing to offer. But how come science is ever useful? Why here and not there? This they cannot possibly tell.

The militant scientists explained this: the Book of Nature is for Man to read, and can be read once the key to it is found. When this (originally cabbalistic) idea seemed too simplistic and unsatisfactory, a more rational and specific argument was found. Take a theory that has been tested and withstood the tests well, says Hermann von Helmholz; its predictions came true either because it is true, or due to some accident. Given these two options, most probably the first is the correct one. It is obvious that Helmholz's argument amply justifies the application, in practical affairs, of any successful scientific theory. Not so the pragmatists: they acknowledge the practical success once it has been achieved. Of course, Helmholz takes the predictive success of science as given and explains the success we have in its practical application, whereas the pragmatist accepts the practical success as the success of scientific predictions, and these one at a time. Taking them one at a time permits success to stop overnight; its very continuation every day is then rendered a miracle. Even the sunrise is a miracle to one who cannot have a theory to explain it, as Whitehead already noticed. How come we have theories, and how come some of them are spectacularly successful in their tests, even Helmholz leaves open; but at least this is accepted as a fact.

Is Helmholz's argument valid? Is it therefore wise to apply every successful theory? No. Helmholz assumes that there are two options given to view a theory that has successfully passed severe tests: one, that the success is due to the theory's truth, and the other, that it is due to an accident. And, he concludes, truth is likelier than a series of systematic accidents. Let us take an example. The sun rises every day. Rather than say that by sheer accident every morning the sun obliges Newton, we can say that Newton is right, hence the sun will rise every day. (Even if Newton is right the sun need not rise forever in the same manner; but this does not matter here.) As it happens, we do not think today that Newton is right. He may be approximately right. This idea we owe to Einstein. It is of immense consequence for technology. It explains, for example, why we do not simply apply every scientific theory, but test its applicability to separate kinds of cases separately: we test separately general facts based on it. We require these tests by law. This fact alone suffices to refute both traditional (Baconian) rationalism and traditional (Duhemian) pragmatism.

Hence, with the transcendence of traditional philosophy of science, a new philosophy of science – and of technology – has to enter the picture and explain, in particular, the interaction between science and technology a bit better than its predecessors, and also, hopefully, help improve the relations between the two. This sounds problematic, since we should notice that any explanation proper should belong to science, since we want our explanations scientific, whereas all practical improvements should belong to technology. What, then, is the task of philosophy here? Perhaps it is to offer explanation schemas and improvement schemas. Yet the schemas need not be meaningless: they can be within language, and hence true or false, and hence capable of being superseded.

5. CONCLUDING REMARKS

The pragmatism of the twentieth century, unlike its medieval predecessor, is secular and liberal. It is popular in scientific and technological circles. It has an impressive, scholarly and technical machinery that goes with it, a theory of truth, a theory of meaning, mathematical techniques, etc. Yet it is, in the last resort, a pathetic defeatism. A great tradition demanded too much in terms of rational justification and so was forced to make do with less. Pragmatists are willing to give up too much. Since pragmatism is a liberal mixture, it is hard to criticize it for what it claims – at least as long as it readily adds to its claims anything which turns up as feasible. For example, though it is fideist, it allows science and even scientific rationalism on fideist grounds. It is better, therefore, to give up the struggle and simply ignore pragmatism. Anyone who insists on pragmatism despite all that has been said thus far against it is simply not very interested in criticism and has the right, in a free society, to have it his own way. The same goes for anyone still sticking to classical rationalism, especially since with little effort anyone can find the way from classical rationalism to rationalistic pragmatism.

The reason is simple: both want to avoid skepticism – because doubt paralyzes. But what is the status of this very claim? Is it scientifically proven that doubt paralyzes? Is it pragmatic to endorse this claim as an article of faith? It is, of course, a patent fact that some people are professed skeptics, others professed anti-skeptics; some in permanent doubt, some oozing self-confidence; some are active, others are helpless: these three polarities produce eight categories. It is a patent fact that examples are easily found for each of these eight categories. The claim that skepticism paralyzes is empirically refuted. Vigorous philosophy of science and technology is therefore quite possible while all doubts are admitted: activity and open-mindedness can go very well together.

CHAPTER 8

CONTEMPLATIVE PHILOSOPHY AND ITS
IMPLICATIONS FOR TECHNOLOGY

It is hard to say what contemplative philosophy is. We had a few occasions to notice that oriental philosophy is practically oriented yet goes under the title of contemplative philosophy because it recommends meditation as a means for the attainment of peace of mind. No Western philosopher has ever gone in this direction to such an extreme, except perhaps Baruch Spinoza. He, too, deemed his view chiefly practical (and hence the title of his most contemplative work: *Ethics*, meaning, a practical guide to life). His maxim, indeed, was that nothing is more enjoyable than learning; the scholar needs nothing but his peace of mind, popular neglect, and his piece of bread to be the happiest of all humans.

Contemplative philosophy, then, cares little about those ends that the vulgar identify as the aim of technology: comfort. This is not to say that the contemplative philosopher cannot value or contemplate technology. He can value technology as a means of improving man's lot, as beautiful and as challenging. He can contemplate, of course, any item which challenges his wits – even the details of a piece of machinery; but this will not count as his activity as a contemplative philosopher. As a contemplative philosopher, he can contemplate the proper ends for technology to serve, the impact of technology on culture, and the nature of the world – physical, biological, social – which makes technology at all possible, or what makes it what we know it to be. Let us begin with the last point: how is technology possible? Why is it different from pure science?

1. SCIENCE AND RATIONAL TECHNOLOGY

The starting point is for a contemplative thinker to contemplate the activity of active people in general. Of course, no person is purely contemplative, except, perhaps, the conscious individual who is kept alive by the modern machinery available only at the most technologically advanced hospitals, though some sort of helpless scholars try hard to approach it; and no person is purely active, except, perhaps the robotized individual of science fiction, though some sort of oppressive beehive regimes may approach this limit.

152

Nevertheless, men of action may thoughtlessly pretend to be robots, and contemplative people may pretend to be free-floating spirits, to use the self-descriptive phrase of the sociologist of knowledge Karl Mannheim (*Ideology and Utopia*). Fictitious as the two extremes are, it is advisable, as a general rule, to idealize – to start from a fictitious simplified extreme, and if the examination is interesting, pursue it, and then correct the outcome if it calls for a correction due to the oversimplification.

How will the contemplative contemplate the active? He may look at the active as a part of that turbulent endless pointless process of blind forces clashing and gnashing and pushing and pulling each other hither and thither. He may follow Plato and Spinoza and assume that only a thinking spirit can put purpose into all this senseless whirling, and only putting sense into the motion of matter renders that motion action proper, so that only the contemplative spirit is truly active. Here we have two extremes that are hardly different from each other in content – they differ, if at all, only in tone. The two ideas, then, are one: true action emerges from contemplation. This idea, it should be observed, is both oriental and Western. Immanuel Kant found it highly disturbing, and on two counts, the physical and the mental. He deemed physics as laws which fully determine the conduct of the material world, thus leaving no room for action, for the intervention of the spirit in the world of matter: there is, then, no place for values in the world of facts. Also, he knew, no matter how elaborate, detailed, or forceful is the life of a spirit, the gulf between it and the world of matter need not be bridged. And he did not know how and when and why the bridging between spirit and matter takes place.

Enter Darwin. Darwin thought that a picture of the emergence of humanity out of the lower orders may be a picture without the gulf between matter and spirit. There is no doubt, now that Darwin's early notebooks are published, that he was philosophically and metaphysically very ambitious and hoped his views would render quite out-of-date all the traditional philosophy of his day, all of which was but an elaboration on the central ideas of John Locke.

Darwin's program was never executed. His own effort at executing it rested on a faulty program: he wished to prevent Man from splitting into body and mind, by showing evolution to be continuous. The continuity, however, is only intuitively an answer, since it does not logically preclude a split: a split can come at a certain point as a result of a continuous process. (When a piece of matter is torn, this is what happens: the distance between two arrays of molecules increases continuously, until it exceeds the range of molecular forces.) Moreover, the continuity theory of Darwin was refuted by the discovery of mutations (= discontinuous changes), first by geologists and then by neo-Mendelian geneticists.

Yet the very Darwinian version of the idea of the evolution of the human mind, the idea that the mind is a part of nature, had an immense impact on contemplative philosophy. Especially the idea that the evolution of the mind is concurrent or even identical with the evolution of the brain became a powerful substitute to the metaphysical theory that the mind is identical with the brain. To clinch matters, the theory of the evolution of the brain as concurrent with the evolution of the hand made this amply clear and tied together the emergence of the contemplative human with the emergence of the active human: though they are different (especially when viewed in the extremist simple manner described above), they evolve as twins.

The questions – how is technology possible? and how do science and technology differ? – to which the present discussion is devoted, have been totally transformed by Darwin, even though his idea was only a sketch of a theory, a mere program. For, to begin with, we developed a picture of the contemplative human and asked how he can act? The image of contemplative human has changed, and so, now we ask how did the contemplative and the active drift apart one day? The answer to this new Darwinian question is rooted in the theory of the division of labor, and the division of labor was evidently a very useful means of survival in a hostile environment.

How is technology possible? To narrow this question down, let us ask, How is rational technology possible? Even this question is too broad, since there are so many implementations of rationally scientific ideas to given ends, most of which are unforeseeable (or else they cannot be patented). Let us ask, then, How can the ideas of science be at all technologically useful? This is the best formulation of the question to date.

The best answer still is that of classical Baconian rationalism. Scientific theories are reliable, and hence, forecasts based on them are reliable; these forecasts are conditional, and technology meets the conditions specified in the desired forecasts so as to make then scientific predictions.

This is the whole of classical, Baconian rationalistic philosophy of technology. It is self-contained, simple, quite unproblematic, and explains the fact that prior to the Einsteinian revolution there was no literature to speak of on the topic at hand – the philosophy of technology. The philosophy of technology was part and parcel of the philosophy of science: science prescribes rational belief and rational action exploits rational belief to attain given goals. The given goals, incidentally, were sanctified by Nature –

especially after they were happily enhanced by Darwinism. Classical rationalist political philosophy (Telesio, Hobbes, Spinoza) viewed survival as the
supreme natural end; the procurement of food and shelter and the possibility
of reproduction and of child-rearing to the age of puberty became part of the
natural end. Darwin endorsed and enhanced these assumptions. With the
discovery of ethology, i.e., animal psychology and sociology, the broad
outline of the philosophy of David Hume, Adam Smith, and Jeremy
Bentham was amply vindicated within a scientific Darwinian framework. (It
was a mock-Darwinian, pseudo-scientific framework, but let us stay with
impressions.) It seemed too obvious and imposing a framework: human
nature requires food, shelter, friends, courtship, male competition over
female plus female submission (the sexist origins of contemporary ethology
are showing) and all that it is required to explain is human society, with its
contemplative and active aspects: the two combine in a Marxist (or post-
Marxist or pseudo-Marxist) theory of praxis.

The classical Baconian rationalist philosophy of technology reduced to
classical Baconian rationalist philosophy of science on the supposition that
it is rational to implement one's rational beliefs in one's effort to attain one's
goals. The goals, then, are given: survival. Let us examine its two central
theses, first that science offers rational beliefs, and second that it is rational
to act in the light of our own beliefs.

Here Darwinism seems to do more then merely enhance and entrench
classical Baconian rationalism. It seems we have found in Darwin the long
expected rebuttal of skepticism and its threat to science as a system of
rational beliefs and rational actions. For Darwinism may apply to the
theories we believe and use in action: only the fittest theory survives. Faith
and implementation of science bespeaks survival. (This view was recently
forcefully restated by Popper and is now ascribed to him, rather than to the
pragmatist John Dewey or to his predecessors Spencer, Mach, and
Boltzmann.)

Darwinism can, at most, decide that of a set of theories competing for the
same econiche only the one which is fittest survives if competition is fierce
enough. But what is the econiche of a theory and what is fierce competition
for it? We have to know this, since different econiches make different
competitors win in different ways. Moreover, there is the conflict between
the long-run survival and the short-run survival, so that the victor in the short
run may be destroyed, and only if the vanquished is not destroyed in the
short run will the species survive. We do have instances for all this – both
from biology (E. B. Ford's study of butterflies) and from technology (upright

looms won in the short run and blocked progress in weaving technology until horizontal looms won the day).

The Darwinian answer to the skeptic, then, is not only pragmatist, it is forced to identify science with technology. It is thus refuted each time the two diverge. And they do. The students of elasticity, hydrodynamics, and similar fields happily developed the continuity theory quite regardless of the many discontinuities observed in physics: they simply declared continuity theory applied mathematics, engineering, or anything else, so as to deny that they were violating the rationality of modern physics. Elasticity, then, is proof for the divergence of science and technology; so are those parts of science which as yet have no conceivable applications.

The people who developed the continuity theory for technological purposes, were of course, right. The question remains: what enables such people to apply their ideas rationally to practical matters?

This criticism indicates that not all rational technology is the application of rational beliefs. A technology may be the application of views not held by anyone. Pragmatists vindicate Newton by the claim that his ideas are still applied by technologists today. This vindicates Newton either on the view that we apply our beliefs, so that we still believe Newton (rationally), or on the view that science is not belief but technology. The mock-Darwinians have tried to unify the two in the pragmatic theory of belief: people survive best when they believe what is most useful to them. Yet we know, every time we pretend (or tell a lie), that it is useful to deviate from one's beliefs. The continuity theory of elasticity was such a pretense.

Also our question has radically changed its character by now. The question was, how can science be technologically useful? The answer identified rational technology with applied science. This identification is refuted. Science gave up Newton in favor of Einstein and of quantum theory, but technology still hangs on to Newton. Hence the question either limits itself to that part of rational technology which is scientific in some very strict sense, or to that part of rational technology which is scientific in a broad sense, or we declare all rational technology in some sense scientific. It matters little whether we take the question this or that way – indeed, we may take it in all three ways, since each way shows a different legitimate and interesting aspect of the problem. And since tradition breaks down here, we are at liberty to view all rational technology as scientific or only part of it.

When we view all rational technology scientific, all we do is stress the fact that the knowledge used by technologists for which scientific status is claimed is largely repeatable observations. We may use hypotheses and complex numbers and rules of thumb in simplifying our computations, all in an effort to link a description to a prediction; but the claim, of course, remains: under the repeated conditions described the prediction repeatedly obtains.

Why, then, do we trust these claims for repeatability?

This is the problem of induction, the problem, how can the skeptic's criticism be met? as translated from the philosophy of science to the philosophy of technology. This way science proper loses all its claim for reliability and thus for science the need to answer the skeptic is overcome – not satisfied but null and void. The need to answer the skeptic in practical life remains: how is rational technology at all reliable?

We may now consider our question in the narrow, strict scientific sense: how is science-based technology at all possible? This is no less interesting. No doubt science does contribute to technology: the facts of science, the observation reports made by pure experimental scientists, are repeatable and at times have desired technological implications. How come? There is no general answer to this question.

Taking the question in the broad scientific sense seems the hardest. Strictly, only current scientific factual observations are what current science contributes to technology, and theories are only means of technological forecasts. Broadly, at times the theories used in technology are contemporary, at times they are out-of-date. At times technologists prefer current theories, at times outdated ones. Why? There is and can be no general answer to this question, only a simple criterion: we use the most handy theory available. Which it is depends on our aims and our theories and our means of computation and more.

2. REPEATABILITY VERSUS RELIABILITY

Many philosophers of science present the process of induction as that of generalizing from a series of experiences. The problem of induction becomes this way the problem of how many experiences, what kind of experiences, or under what conditions experiences permit or guarantee the conclusions of a generalization? Or, how firm, probable or credible is such a conclusion and why? These questions are all not to the point, since science begins with generalizations. Generalizations are the given which science attempts to explain. The first question to ask concerning a given generalization is, Is it true? And in order to find out one tests it. Unfortunately, tests only refute, never verify. So when we refute an observations statement we correct it as best we can, and when we find a correction we take it as given and test it. When we cannot refute, says Imre Lakatos following Popper's footsteps, we explain. The explanation often is at variance with what has been set to explain, says Popper in the footsteps of Einstein. Thus, Newton explains not why terrestrial gravity is constant (which is what Galileo has claimed), but why is it so nearly constant. And so, crucial experiment between Newton and Galileo is possible. And, we all know that Newton won. And then Einstein won against Newton. And so the field of observation reports presented as generalizations does increase all the time, constantly modifying old ones and making room for new ones.

How does this whole process take place?

This is the central problem which philosophers these days concern themselves with. The answer to this question can be scientific, metaphysical, or logical. Science does offer partial answers to this question. Yet even were the answer which science should offer tomorrow complete, this answer would not satisfy, since if it is scientific then the explanation of the scientific process is a part of the process to be explained. A logical answer would be fine and solve the problem of induction once and for all and restore (Baconian) rationalism. A logical answer will never be discovered since it is impossible. A metaphysical answer will read something like this – the world is at least partly comprehensible and we humans are endowed with the ability to comprehend it. Bacon invoked the authority of the Bible: God promised the sons of Noah domination over Nature, and true domination is possible only through comprehension(!), so that God has promised science success. Modern Baconians argue similarly except that they invoke Darwin ex machina.

Apart from all this, no one stops us from developing a science of repeatability and reliability. Except that a theory of reliability need not be reliable: it must explain. And a theory of repeatability does not need for the observation reports it uses anything more than the claim for repeatability. This is the best avenue for research.

To open this avenue and keep it open, all we need is to notice the difference between claims for repeatability and for reliability.

Let us observe a factory which is quite reliable. What makes it reliable is its good quality control. Improved quality control tests more characteristics with higher standards and – most significant – misses fewer defective items than before. This raises production costs. The nearer to zero the desired percentage of poor quality products put out, the higher the cost of quality control; so much so, that there is an obvious cut-off point, beyond which the factory must allow for the odd defective product to reach the market; the factory then finds it cheaper to handsomely compensate the consumer who happens to have purchased the defective specimen than to tighten quality control.

Not only can we wonder at the reliability of the product, i.e., of all the generalizations used both in production and in quality control; we can also wonder at the stability of a society which employs standards of reliability, rules for complaints and for compensation, and more.

We thus meet a variety of attitudes to defects – whether of the production process or of the control process. The production engineer is happy if the reliable products constitute some ninety percent of the product, though in some industries it is impossible these days to achieve much more than fifty percent (lasers, computer chips). The production engineer is happy if the quality control mechanism rejects ninety percent of the defective products, or ninety-nine, or whatever is required by current standard. And as long as the standard is met, the production engineer is happy and is not troubled by the defective product.

In other words, in technology success is determined not as full success – this is humanly impossible – but as a level or success declared satisfactory by some standard. In science success is determined in the diametrically opposite way. Science is concerned not with successful predictions but with failed ones. The technologist is happy to let the defective item be rejected; the scientific researcher will consider it a matter of greater interest than the successfully produced item.

(This point is not only overlooked – it is violated. Science and technology these days are presented as one concern, and for pragmatic reasons: they are identified simply because of vested interests that need not be examined here. Consequently, science is presented as the hunt of successful predictions, not of the correction of mistaken ones. God-fearing honest philosophers of science follow the party-line and tell us in all the priestly sincerity they can muster that a hypothesis ninety-nine percent true is fine, that we do not overthrow a hypothesis unless we have a replacement for it so that no hypothesis is ever shown false, and that we always have to believe some theory or another since we always must act.)

We can conclude some obvious corollaries concerning repeatability and reliability and the kind of answers contemplative philosophy can offer to the questions, How is science possible? and from whence comes the reliability of rational technology?

The Darwinian approach can be stripped of the pretensions which contemporary pragmatist fashions wrap it in and can be used to describe an environment that is life-sustaining so that for one reason or another it shows certain regularities. Some of these regularities are rooted in laws of nature; others are mere accidents, and the results of certain accidental characteristics of our life-supporting environment alone. We want to know which is which; we particularly want to know what vital, accidental factors may be drastically altered, so as to try to maintain or replace them, so as to avoid the fate of the dinosaur. Extinction or the threat of extinction must be included in any properly Darwinian picture of humanity. Consequently reliability becomes partly a local matter, not a matter of logic or of metaphysics; it remains for science to examine how much of our safety depends on natural law and how much on local accidents.

Yet Darwinism also says that each stage of our cognition has a previous stage, and the lowest stage is just above the stage of comprehension of a simian. This leads one both to animal psychology and to the simplest levels of human comprehension, those levels of human comprehension which are common to all humans, commonly known as commonsense.

3. SCIENCE AND COMMON SENSE

Contemplative philosophy is the way contemplative people approach the world when moved by the desire to comprehend it. Diverse contemplative philosophies direct attention to different parts of aspects of the world as the worthiest of attention, and offer different intellectual frameworks and different tools for such comprehension. What is comprehension? This is a very hard question which invites conflicting answers. One may regard all this as secondary and approach comprehension commonsensically: We have a common and rather unproblematic sense of what we deem problematic and of when the problem is solved – erroneously or correctly, but solved.

Perhaps the most important, most obvious, most often violated maxim concerning comprehension, or concerning a solution to any problem, is the ancient maxim which may still be ascribed to the modern American philosopher Charles Saunders Pierce: Do not examine the truth or falsity of an irrelevant statement; when a question is asked and elicits a response, the attempt to examine whether the response qualifies as an answer logically precedes the attempt to discover its truth-value (i.e., whether it is true or false): It is unwise to skip the first step. If the response is not an answer, at least it should be relevant to the question.

What is relevance? In logic, relevance is any logical relation other than independence: two views, two statements, are relevant if and only if one follows from another or contradicts it – as such or in view of some background information. If we view relevance this way, then, we are told, comprehension should be viewed as explanation and explanation as de- duction.

This will not do. Some deductions do not count as explanations, especially technological deductions, such as "explaining" the behavior of a gas, a fluid, or an elastic solid, on the assumption that atoms do not exist. Also, explana- tion is often not a deduction of the problematic statement of observed fact but of a slightly different one, the corrected statement of observed fact; not, all bodies fall with a fixed acceleration, but, all bodies fall with a nearly fixed acceleration. Hence a Newtonian explanation of Galileo's Law is not de- duction. With this the relevance of logic to relevance becomes highly questionable.

Is this critique valuable? Does it matter that not all deduction considered seriously is explanatory and not exactly what is to be explained is deduced? No. It only shows that comprehension or relevance cannot be captured by logic alone the way philosophers of science have traditionally described it. But we can easily entrust the question of relevance to common sense.

This is why such great scientists as Albert Einstein and Max Born insisted that science is but an extension of common sense.

This view, however, slightly alters our view of common sense. For, to begin with, we present common sense as the very knowledge, understanding, sense of proportion, and approach to diverse matters, common to all. Now this quality is expandable, but only when the expansion gains immediate general acceptance. Thus, when we say of someone that he is eminently commonsensical, we mean he has more common sense than is common, as evidenced by our seeking of his advice on this account: he has more sense, and when he shares it, it becomes common at once. Good advice is, indeed, obvious after it is given but not before.

Is science the extension of commonsense in this respect? In many cases we say yes, unhesitatingly. Quite a few scientific ideas strike one as so common sense that one wonders why they had to be invented in the first place. History shows that such ideas win immediate acceptance. But not all scientific ideas are accepted without a struggle. Why? Because, the Baconian traditional answer has it, even scientists can be prejudiced. This is true but unsatisfactory: why are scientists ready to accept as obvious one idea and not another?

The answer to this question must be commonsense. The commonsense answer is, indeed, very obvious and – to report an empirical observation – wins immediate acceptance, especially from those who have spent some time in an attempt to solve the problem. The scientific idea that wins immediate acceptance integrates well within an existing intellectual framework. Scientific ideas which refuse to integrate with ease into the existing scientific fabric are revolutionary. Revolutionary ideas, too, may gain quick acceptance, if and when the general public is ready for a revolution and yearns for it (as Thomas S. Kuhn observes). Thus, an idea may win immediate acceptance either if it integrates into the scientific fabric or into the higher-level set of expectations about the fabric itself. But then the lower level, too, must adjust so as to make the revolutionary scientific idea common sense: usually, the new idea replaces old ideas. Hence, common sense alters not only by accretion but also by some improvement through transformation. And with common sense we judge the replacement adequate. How else can we do so? If we have no commonsense criterion to judge the adequacy of the change of common sense by, then common sense lose its independence.

In order to move towards a scientific theory of the applicability of science to technology, then, we may start with two aspects of the situation. First, we may go back to the question: What is scientific comprehension? For we have attacked the question and refuted the received answer to it and argued that common sense takes good care of it. What we may want, now, is to contemplate the way common sense does this. We may then attempt the same with applied science. Or, we may find, as seems rather obvious, that common sense makes at times no distinction between pure and applied science, between comprehension and practice.

For example, the most commonsense idea in both science and technology is to seek quantitative functions and to deem them linear. The idea that there are numerical functions is often commonsensical: every normal person, even with no science and even from an unscientific society, will agree that given a simple and fairly constant fire, and a simple container of water on it, the more water will take longer to boil. On the whole, the idea of linearization is no more nor less than the idea that any curve may be approximated by line segments, and the shorter the segments the better the approximation.

It is most significant to notice that the ideas expressed in the previous paragraph are not known to most people yet they are commonsense. That is to say, in the sense of being commonly held they certainly are not in the least commonsense, yet they are commonsense in the sense that they are readily acceptable once they are grasped.

This kind of idea attains practical acceptability quicker than scientifically sophisticated ones. There is a distinct tendency of technologists – especially those technocratically oriented – to go the other way and seek sophisticated ideas. And, indeed, the other way is at times immensely valuable in that it accelerates, when successful, the rendering of big chunks of sophisticated science and technology into common sense.

Yet once we develop a better eye for the common sense extant in science and a better feel for the urgent need to disseminate the scientific knowledge essential for democratic control, then we will develop a new field of human knowledge, comfortably seated between high powered science and popular science, and repeatedly used for the improvement of the quality of life in democratic society.

Common sense is our ability to apply existing knowledge to diverse situations – including problematic situations, we may remember – in an unproblematic and uncontroverted way. This is not finality. In particular, there can be no finality in common sense since it includes diverse and conflicting ideas, including a vague but generally understood general framework, which is far from being clear and distinct, but without which everything goes to pieces, including the flexibility that is the readiness to change.

Here, then, is what vexes novices in philosophy so much. Common sense takes so much for granted it leaves too little room for exploration. Classical rationalism begins with (Bacon's and Descartes') efforts to start from scratch. This is too incapacitating. If we cannot begin from scratch, then we must begin somewhere. Where? We begin by accepting an intellectual framework. Both Michael Polanyi and Karl Popper, the leading irrationalist and rationalist of the mid-century, agreed on this, and even on its being a matter of common sense. What is the framework? Is it common sense? Is it articulated? Is it rational? Polanyi says the framework is common sense, but common not to all people – only to the experts in the field, whatever the field may be; nor can it spread to all, since it is tacit knowledge: it need not and cannot be articulated. Thomas S. Kuhn added to this his idea of the paradigm (the word means chief example) which is at times tacit knowledge, and at times the framework. For, at times it is the framework which can intimate to us what examples to emulate. Popper, on the contrary, denies the framework any significance except heuristic: one thinker's framework may make him suggest one hypothesis, and another may disagree with him. The common framework, said Popper in his classic *Logik der Forschung* (= logic of inquiry) of 1935, is not common views but common methods: choose the most highly testable hypothesis, test it and try to refute it as best possible, and so on. (Things always get problematic with the so on bit, but that is a different matter.) Popper himself later changed his views, though gradually and without ever taking stock of the change, possibly not even recognizing its mere existence. He said that the way we decide how to test a hypothesis depends on the framework of our common background knowledge, which is a mixed bag. Background knowledge, then, is common sense. Using it, then, is looking at new scientific ideas from the view of what was already popular before. After it gains acceptance, then, it enters common sense. This makes today's common sense the frontiers of science of yesterday. This view of common sense was suggested by James Clerk Maxwell, better known for his equations for electromagnetism.

We have here two variants of the idea of the framework, and in Popper's and Polanyi's views a blend of the two. One variant is of the framework as first principles as applied to scientific research; the other as the common-sense sedimentation of past research. Common sense need not be more sedimentation of past research, however. It can explode with revolutionary new principles, such as Bacon's view of common sense or its application to the social and political sphere; or Romanticism, Marxism, Darwinism, Freudianism; and many other ideas, scientific, unscientific and anti-scientific. Yet, as we cannot operate without common sense and since it is in a constant mess, it is sheer common sense to approve of efforts to improve it somehow. And the easiest is to take the principles within which we operate and see what is their relation both to science and to its sedimentation into common sense. These first principles are usually known as metaphysics.

4. METAPHYSICAL FRAMEWORK

Metaphysical systems, said Bertrand Russell, are visions of the world which, once grasped, seem very simple indeed. Of course, they get less simple when we try to criticize them or even when we try to apply them to diverse cases, including ones that were remote from their originators' intentions. Yet, obviously, some systems are extremely easy to apply. An example would be any magical system one happens to be familiar with or to invent for oneself. Yet the very applicability of a system everywhere is seductive, as Popper has noticed. For an example we may take two competing metaphysical systems of psychology, Freud's and Adler's. Freud saw all action as motivated by the sex drive, Adler by the desire for social acceptability. Now Freud also recognized the desire for social acceptability, but declared that it derived from the sex-drive. And Adler reversed the order. Thus, Freud explained courting as an overture to sexual encounter, and Adler saw sexual encounter as a part of courtship. On the basis of new ethological developments, incidentally, we now reject both views, since today clearly higher animal courtship is viewed as an integral part of the normal sexual encounter (and copulation without courtship is, consequently, a perversion, which Konrad Lorenz, for example, saw among some homosexual birds, which, he reports, keep courtship and copulation quite separate, one homosexual and one heterosexual).

Justificationist philosophers since Bacon were all sensitive to the role that intellectual frameworks play in empirical observations. The question they raised and continue to raise is that of legitimacy. All intellectual frameworks except the true one are not legitimate, and even dangerous, said Bacon, since by a vicious circle the framework and the items within it keep supporting each other. Hence, Bacon demanded that science operate without any intellectual framework first, and let it evolve from the facts slowly and in small steps: the framework should appear last and have the status, he added, of scientific metaphysics.

It does not. All attempts to describe facts without an intellectual framework, pure facts, neat facts, data untarnished by preconceived notions – they all failed. The latest attempts were all destroyed by the fact that an observer reports his observations within a space-time coordinate system. Systems, then, precede reports. This is the point made by Galileo, Kant and many other philosophers and perception theorists. It was, indeed, Descartes' wish to establish a framework for science, a metaphysics to make science possible. The question was, how could this metaphysics be proven. Descartes said that clarity and distinctness are proof enough. Kant said that since without a framework science is impossible, a framework must exist. This is questionable: do we have science? Most people respond with, Of course we have science, since what else do they teach in the university? But let us remember that what was taught in the universities since their foundation in the Middle Ages or even in the last century is not what today would be at all acceptable to us. If science is final, we have no finality. And if science is so defined as to include defunct ideas, the justificationist philosophies – the likes of what Descartes or Kant have offered – are all left behind, since their justification itself is final or else questionable.

The last resort was an attempt to have only space and time as a framework, and rely on Euclidean geometry. The discovery of non-Euclidean geometry was unpleasant but not fatal for this program. Einstein put an end to it, more with his deviation from the common traditional concept of universal time than with his deviation from Euclideanism, and his deviation from Euclideanism was more shocking as a possible scientific theory than as the allegedly true one: Einstein was a skeptic to the last, and to the last he found his own theories wanting since not in a sufficient conformity to his metaphysical ideal. He rejected the idea that both space-time and matter are independently assumed to exist (and this is what general relativity assumes when it equates some of the properties of space-time with some of the properties of matter-energy): this idea was to him intolerable except as a mere stop-gap.

The question remains, how do we choose between competing frameworks? Whatever else we say, we may notice that either the choice is utterly arbitrary – fideism – or fully reasoned – justificationism! What other option is there? There are two other options; and the same technique may lead to both: when all answers to a question are troublesome, we may well want to see why, before going any further.

It seems an obvious fact that our very question (how do we choose a framework?) is justificationist; so much so that it forces us to be justificationists on pain of becoming capricious irrationalists as the only alternative. Our question is justificationist also in its presupposition that it is at all given to us to approach the choice of a framework with no presuppositions and no prior framework. Our question may read, If I were a tabula rasa – a clean slate – where would I begin and how would I judge where to begin? But none of us is ever a tabula rasa, and so the question should better read, How best can we use whatever ideas and tools we have now in the choice of our next intellectual framework?

Now the new framework, or the presuppositions it includes, its metaphysics, is not the same as background knowledge, since background knowledge already exists and is used as a means for the choice of that metaphysics. And background knowledge includes, of course, earlier metaphysics, which contradicts currently competing metaphysical frameworks.

Can background knowledge, a mixture of old ideas, help us choose a new system? Will not the old impede the growth of the new? Fideists will dismiss this discussion. Never mind how the new framework is achieved, they would say, as long as it is consistent, workable, and accepted. It is impossible to judge the new by the old, except negatively, they will add, and prove that by logic. Hence, they propose, let us not examine one system by its background, nor the other way around. This is the view of Duhem and of Evans-Pritchard, currently publicly known by the ignorant as the Kuhn–Feyerabend incommensurability thesis. Facts refute this theory. Judaism and Catholicism say: prefer me over all else under any condition. And so they lived side by side for ages. Newtonianism did not say so. And it made a great difference. One century ago practically all physicists were orthodox Newtonians. The representatives of the scientific community then repeatedly said that Newtonianism will never be superseded or overthrown; they even went so far as to say that Newtonian mechanics will never even be explained (Herschel; Whewell; Mill; Poincaré). Yet, within one or two decades early in the twentieth century, almost all physicists changed their views. How? Why?

Some say that there was a crucial experiment between Newtonian and Einsteinian astronomy in 1919. Others say that relativity won on so many fronts. Still others see here a shift in intellectual frameworks. Perhaps they all make some important contribution to the true explanation. Yet it is best to reject them all and say, instead, it was none of these and not a mere combination of these but mere common sense. To avoid sounding like Michael Polanyi and Thomas Kuhn, let us render this distinct from their irrationalism and violate their claim that it is impossible to articulate the idea of how exactly such transitions are made. Here, then, is an attempt to articulate the view that the major transitions are effected by common sense.

Not only is the question, How do we choose between frameworks? erroneous since we are not clean slates. This was corrected by asking, given our background knowledge, How do we choose between frameworks? The question, however, should undergo a much deeper revision: what do we choose, and what can we choose, and how can we improve our mode of choice? For, one may contend, we do not choose basics of faith: they are given, and we do not always choose as best we can – at times only we can improve our ways of thinking.

The claim that we do not choose basics of faith has been made repeatedly. If a prince commanded me to believe this or that, said Robert Boyle (*Occasional Reflections*), I would not know how to obey him. And Spinoza based much of his political philosophy on the impossibility of forcing people to believe. And the greatest pragmatist philosopher, Charles Saunders Peirce, whose major writings are two papers, one on clear thinking and one on the fixation of beliefs, admitted in a private letter to his friend William James that we cannot control our beliefs. You do not know how these things happen says George Orwell's *Clergyman's Daughter* out of the depth of her heart and experience: one day you wake up and simply know that your faith has gone and there is nothing you can do about it.

What, then, do research workers do? They do not always choose first and try out later; at times they try out first. Nor do they always accept the verdict of their attempts: they can try again. Erwin Schrödinger deemed the Einsteinian framework better than the Newtonian framework in which he constructed his celebrated equation for the electron; he even believed it was true. Indeed, before he developed the famous Schrödinger equation he developed a similar one, within Einstein's system (later rediscovered by Klein and Gordon). It did not suit his purposes, so he tried to use the simpler, Newtonian framework instead: he always tried to apply his ideas within different frameworks. When Einstein developed, at the end of his career, his generalized or unified field theory, Schrödinger likewise did not ask, Do I believe it? Do I choose to believe it? or any such question: rather, he tried to use it and develop it.

This suggests, then, that science has to do with what we happen to think may be true, or nearer to the truth, or even pointing the way to the truth, without suggesting that either experience alone or the framework alone, or even both together, are always capable of deciding matters. This keeps things nicely open, taking best account of what little we know and of the fact that so much we do not know. This is sheer common sense. For, it is not common sense to decide what to believe when we do not know but must decide: when we must decide, we decide, but we need not believe. And then we may analyze common sense and see that in all unproblematic situations we are all too credulous, that when a situation is problematic we try to extricate ourselves from a problem this way and that way. At times we work on tacit suppositions (as Polanyi has noticed), which we admittedly cannot as yet articulate and which we hope to articulate later. At other times we devise a strategy and discuss it at length (at times the length of a few generations at a time) and then try to execute the program. The strategy may be to develop a metaphysics or to develop a way to apply it to a given set of problems. How all this affects our beliefs can be studied too.

What complicates matters is that pure science is the opposite to applied science and technology, yet research is an activity to which we can apply all the tools we have at our disposal: for, the technology of research is a cross-breed and so a new breed.

This examination of metaphysical frameworks is thus pluralistic, but at the cost of leaving the questions of belief and of truth in mid-air. Clearly the very search for truth is not, as classical rationalists asserted, a matter of proper belief. Religious researchers and secular ones can work together. Researchers can work within frameworks they consider false. Yet their end is the truth; they act on the supposition that they may further that end. They also do have beliefs and these are at times highly relevant. Hence, we can only study specific cases and kinds of specific cases; but we can do so only in a given framework, which is, of course, question-begging.

Since belief is private, may we also assume that it has no more than a heuristic significance? Or are we obliged to accept – i.e., believe – on each given issue the one hypothesis concerning it which has best withstood empirical testing? Perhaps we can examine this matter itself empirically – by watching faith in action. Where does belief spring into action? Do we act on our beliefs in daily life? Do we do so in research? Where in research does belief spring into action, where in daily life? (The same questions concerning belief can be asked of rational belief or credibility, of course.)

This barrage of questions is petrifying. Polanyi takes them with ease, taking up, in response, one broad and important fact. In research, he claims as a matter of fact, faith enters through education and through the resultant coordination between research scientists. This is dreadful: this way Polanyi takes scientific education to be indoctrination. Admittedly, in fact too much of science education in college is indoctrination. But at least the novice in science must be told of defunct theories to be taken seriously, of alternatives not to be dismissed out of hand, and of problems concerning the most firmly received opinions of the scientific community. And so, sooner or later a well-educated scientist may become autonomous, develop his own mind, qualify for independent research. Sooner or later he may conduct interesting research and he may be noticed by his peers. As Einstein has noted, science grows despite the dogmatism of science education, though he never tired of preaching an improvement in scientific education.

In daily life there are instituted opinions. These need not be received. Indeed, while Aristotle's astronomical opinion was received and dogmatically transmitted in universities, first Ptolemy and then Copernicus were instituted by the church of Rome for purely practical purposes. How did Bellarmino decide to institute Copernicanism at the time he viewed it as false? On what ground? How do views such as Newtonian mechanics become instituted in naval academies? This is the new variant of the old problems. And once we handle it well we can ask, how did the naval academies, military and civil, contribute to the rise of Newtonian metaphysics, the scientific attitude, and the disposition to welcome and try out technological innovations, and how did these, in their turn, get instituted?

5. CONCLUDING REMARKS

The history of modern physics presents us with a few intellectual frameworks. The first is an outgrowth of ancient – indeed Presocratic – ideas, Pythagorean and atomistic, which via the cabbalah evolved into mechanism. The second was Newtonian. The third was field theory. And recent physics is struggling toward finding a fourth. Einstein has effected both a scientific and a metaphysical revolution. His scientific revolution was viewed as a proof of pragmatism. When the empiricists emerged from their shock they declared scientific theories not certain but probable and Einstein's theory more probable than Newton's. At the same time the feeling strengthened that something more is required of the philosopher of science. It was felt that the overthrow of Newton cannot be viewed the same way as the overthrow of Aristotle. Lord Rutherford declared it silly to think Einstein overthrew Newton, and he put it in a wording that is very obscure; yet Einstein is reported to have used the same wording (in Chaplin's autobiography; Chaplin could not possibly have read the Rutherford original). Werner Heisenberg, the celebrated scientist who also was a poor philosopher, has presented a variant of instrumentalism which he called the theory of closed theories. It says that no overthrow of a scientific theory is possible. This claim is contrary to known facts and based on a confusion of different kinds of overthrow, within science and without. Popper's demarcation of scientific theories as refutable explanations plus his articulation of Einstein's theory of degrees of approximation to the truth are now seen by an increasing number of people as an adequate view of the matter. For the time being.

Yet Einstein said more. He said it was possible to modify Newtonian mechanics so as to close the gap between theory and experiment and even to do so adequately, yet he was never interested in such a project since it looked to him artificial. Karl Popper viewed this unease as the reluctance to accept ad hoc hypotheses – this reluctance as rooted in the desire for simplicity, and the desire for simplicity as the search for the most highly testable theory possible. Thus, there is a special status to Popper's requirement for the highest degree of testability, atop his demarcation of science from non-science by the claim that science equals testable explanation. That requirement explains Einstein's reluctance to tinker with Newtonian mechanics without making mention of Einstein's metaphysics, his general field theory, the intellectual framework within which he tried to work.

Here Popper is in error and confusion. The demand for the highest testability in research leaves no room for pluralism in research. High testability is required not in science, pure or applied, not in research, but in practical technology, particularly in nuclear technology and in space technology. In technology the modifications (for example, of Newtonian mechanics) which researchers (for example, Einstein) consider distasteful and arbitrary, are developed by applied scientists, to be tested as severely as possible and applied by technologists. Tests in technology are different from tests in science: they are attempts to refute claims for safety, not descriptions of any old facts of nature.

CHAPTER 9

SOCIAL PHILOSOPHY AND ITS IMPLICATIONS
FOR TECHNOLOGY

What the social philosophers of the traditional Enlightenment Movement, as well as those of the Romantic Movement, said about society is best presented as intellectual frameworks for social sciences. Yet one must beware of one trap if one wishes to avoid gross obvious errors: it is one thing to have a framework, and even develop one's ideas within it, and quite another to stay within it systematically – which is neither desirable nor possible. Some fashionable philosophers take the undesirable and impossible to be both desirable and matter of fact (Kuhn, Lakatos).

When we come to the study of society, the deviations from the intellectual framework become very significant for two reasons. First, the framework in the social sciences is not as good as in the natural sciences. Second, in both fields practical matters press and force people to deviate from their framework, but in social affairs this fact is both more conspicuous and more problematic. The problematic aspect is also enhanced by much bigotry.

The traditional Enlightenment philosophers were *reductionists*, i.e. they believed that all science is ultimately physics. Of course biology did not stay with physics alone, nor psychology with physiology. Yet sociology and politics were repeatedly pushed towards the framework of psychology, which was a straightjacket. And reductionism became, and still is, subject to much bigotry. Thus, social affairs and political affairs were either left outside the field of applied science or highly constrained by it. And then the Romantics tried to reduce psychology and sociology to politics, and with similar constraints. This was both a blessing and a curse.

1. REDUCTIONISM IN ACTION

Reductionism is attributable to Auguste Comte, but is traditional and is directly linked to classical scientific individualism. The Latin individual = the Greek atom = the English indivisible. It is also linked to the scientific claim that human nature is one and the same everywhere, and to the inductivist claim that we can see people, but not societies. More profoundly it is linked with the worship of Mother Nature, common to the Enlightenment and the Romantic Movements alike. For both atomism and individualism shared the hostility to animism, a hostility which led to nature worship in the following and very odd way.

The main question of the philosophy of Man are two: is Man inherently 'good'? and does Man have a 'divine' soul? It is important that the question about goodness is about inherent goodness, not perceived goodness, because we have perceived both 'goodness' and 'evil' in human conduct: likewise that the question pertains to 'divine' souls, not to souls in the sense that people are perceived to be the most intelligent of animals. One might expect to find in literature the view that Man is inherently 'good' because he has a 'divine' soul, or inherently 'evil' because he lacks it. The facts are different. The literature includes works in the traditional medieval philosophy of the Jewish and Greek religious and intellectual heritage and rebellious works of the Enlightenment with reversed judgment on both questions. The medieval philosophy said that Man has a 'divine' soul yet is inherently 'evil': it saw all natural drives as 'evil'. The rationalists of the Enlightenment Movement denied Man his 'divine' soul and endowed him with 'Natural Goodness': it saw Nature as meaningless and indifferent and so it considered the set of Man's natural drives as the sole guide for action and the sole criterion for goodness. Philosophers like Hume and Kant who insisted on the distinction between fact and value, had nevertheless no hesitation in ascribing intentions to Nature and concluding that whatever does not suit Her design cannot exist.

The metaphysics of the Enlightenment Movement was introduced under the guise of methodology: the theory of 'divine soul' was deemed metaphysical and rejected as often as people dared contradict current religious doctrine – which was done openly and systematically only after the advent of Darwinism. Moreover, even sincerely religious thinkers who profoundly rejected reductionist metaphysics were well disposed towards the adoption of reductionism as a methodology. They said that this reductionism renders science devoid of all meaning. Yet life is full of meaning, exhibited in every artistic and in every religious experience. These experiences call for an alternative, extra-scientific methodology, which leads straight to God and to the supposition that our souls are of divine origin. The trouble is, God had to be either the universal God of the rationalists of the Enlightenment Movement or the 'parochial God' of the obscurantists. The 'divine' soul could return to the universal deity or go to a parochial life in heaven or hell. The idea was still hardly discussed that the 'parochial' God is only a metaphor for the universal God and so does not call for obscurantism. This idea is today common, but alas! it is only a new subterfuge of parochialism.

Reductionism calls for a lengthy and detailed discussion.[5] Here we should rather explore some significant corollaries for both social science and social technology. The scientific tradition was a companion to the liberal tradition, and both of them, more or less, to parliamentary democracy: this is said repeatedly. Let us accept this questionable claim for a fact. This fact may indicate that there are some deep-seated connections between science, liberalism, and democracy: or it may be sheer accident: or it may be due to the belief scientists have in liberal democracy. Which of these three options is true? The question is worth exploring, especially in the face of the rise of technocracy, and particularly in view of the urgent tasks of exporting both democracy and technology.

Let us take toleration first. Consider religious toleration, then. The fact that reductionism is a metaphysics thinly disguised as a methodology will not be discussed here. The fact that it prescribes one or two possible attitudes to established religion has to be discussed, since reductionism thus limits the whole range of attitudes that are related to religion – and practically all social and individual matters are related to religion. Yet reluctance to discuss the relations between science and religion blocks this discussion. The result is the confusion that inevitably accompanies veiled discussion. The inability to discuss matters freely in eighteenth century Europe during the Age of Reason (David Hume and Denis Diderot for example spoke quite frankly and almost freely, but they were the exception rather than the rule) and the immense gulf between theory and practice on the Continent of Europe and even in Britain, caused a strange picture to emerge. Reductionism, which was extremely popular as a metaphysics, particularly because of its opposition to established religion, was of great practical importance because it celebrated Nature. It took hedonism for granted, so that the relief of human suffering through knowledge became a major human purpose. Expressions such as 'benefactor of mankind' (taken from Bacon's scientific utopia, his *New Atlantis*) were not used lightly: The idea of the salvation of mankind on earth through knowledge was the religion of the Enlightenment, regardless of the question of whether it was deemed irreconcilable with Christianity. This question depended, more than on anything else, on the possibility of reconciling Christianity with the religion of the nature-worshipper – on its avowed uncompromising advocacy of the doctrine of the natural goodness of Man, namely hedonism. But, whether this reconciliation can be effected or not – and obviously it cannot – the question is, can toleration be effected? This is not so easy to answer.

There are two versions of hedonism, both of which played a profoundly important role in the rise of scientific-technological society, not to mention the amalgamation of science and technology that recently has been achieved.

First there is the demand to tolerate and accept each individual's preferences, such as they are. This is the doctrine of tastes as given, as heterogenous or exogenous, which is the backbone of the theory of consumers' behavior and thus of the theory of the market mechanism. There is no doubt that this theory is obviously deficient and only a very strong tradition could prevent the economists of the classical or monetarist persuasion from observing that were tastes capriciously variable and should they vary often enough, the market could not possibly achieve any stability, let alone equilibrium. There is not enough study of how stable tastes need be to maintain the system. They have to vary in periods much larger than the market's time-lag. The most important deviation from both the claim that tastes are exogenous and that the market mechanism is best not interfered with, or at least only monetarily, is John Maynard Keynes's theory of liquidity preference: when the stock market reaches a certain level of instability, no matter how fast share-prices fall, there is still no buyer to stop the fall. This thesis is the central deviation from classical economics known as the Keynesian revolution.

Second there is the reductionist hedonistic thesis that all humans share the same inherent tastes. This second thesis may be a corrective to the first, since some of our revealed tastes may be more due to corruption than to human nature.

This is a very important point. The Marquis de Sade claimed to have refuted the philosophy of the Enlightenment by arguing that in fact some people enjoy hurting others. The Enlightenment Movement ignored this claim. It had enough trouble handling the fact that people are constantly in conflicts of interest. The spokesmen of the Enlightenment said that science and technology will create such a plenty that there will be little possibility of conflict, and sweet reasonableness will iron out what conflicts may be left. But this is impossible if the conflict is rooted in the desire to hurt rather than in hurt unwittingly resulting from innocuous desire. To this the only answer was that the desire to hurt is quite 'perverse', quite alien to human nature. Economists are aware of this point and they do make it explicit quite often in the form of the hypothesis that people are not malicious and do not intentionally prefer to see the other fellow worse off.

The importance of this discussion can be seen from the fact that the greatest critic of the Enlightenment, Georg Wilhelm Friedrich Hegel, criticized just this hypothesis and replaced it by his celebrated doctrine of the master and slave: every two people have the need to determine who between them is the master. The sadist character of this hypothesis was made explicit by the Hegelian Jean Paul Sartre. This view of human nature is the polar opposite of the doctrine of Jeremy Bentham, according to which all men by nature wish to be friends with each other. Indeed, Bentham saw here a parallel to Newton's law of universal gravitation and considered himself the Newton of the social sciences, all observed facts to the contrary notwithstanding. Nor did matters stop here. Reductionism does push us towards finding out people's inherent wishes or needs or desires or drives. And Darwinism greatly legitimized this by endorsing the classical (Telesio, Hobbes, Spinoza) claim that survival is the chief end of members of any species or else it be extinct, and survival requires food, shelter, reproduction. Ignoring food as rather unproblematic (which shows how parochial our discussion is, since hunger is so common), we are now brought again into the debate between Freud, who saw the sex motive as the prime motive, and Adler, who saw the desire for social acceptability as primary and thus the need for security as the prime motive. The discussion was not about the need for sex or for security, since both needs are generally admitted; the discussion was about which of the two is primary. Freud said that one wants acceptability in order to achieve sexual gratification. Adler said that it is the other way around. And, to repeat, both views were superseded by Lorenz who views sex as a combination of courtship, copulation and homemaking in all animals which court to any degree. The debate is significant only because reductionism is influential in scientific research. Freud and Adler disagreed not only about principle, but they tended to ascribe motives to their patients in accord with their principles. This is a traditional error; one may wonder how the whole tradition of Enlightenment could tolerate it, and how it could survive to this very day. For if one wants to follow reductionism rigorously, then one must first establish physics, develop chemistry on the foundation physics offers, develop biology only on the foundation of physics and chemistry and so one. Once we agree that reduction can wait, that we can develop one branch before the branch which precedes it is completed, the debate loses its immediate import. The methodological principle of reductionism is often viewed as Occam's razor which shaves off entities that can be explained. But there is no rule as to how long a beard we may grow before we decide to shave it off. This is the logical flaw in all the reductionist discussions. Once we make reductionism a matter of principle only, then the hedonistic principle referring to people's present preferences gains weight, while the hedonistic principle which refers to the preferences inherent in the human race may be postponed indefinitely altogether. Then even the principle referring to present preferences becomes inapplicable: people do have conflicts of interest and do exhibit desires to be one-up on their neighbors and some of them simply prefer marauding, pilfering and theft to honest work, enjoy hurting their neighbors, beating up their own wives, and so on. Without the hedonism-in-principle which corrects the hedonism-in-practice, the latter too is inoperable. Hence the vague allusion to preference. Economists speak of preferences today some-times as they are revealed in the market and sometimes as they presumably should be in accord with human nature. This shiftiness is very central: men are not good, but their 'natures' are good, and so their revealed deviation from goodness is a technical impediment to be overcome by increasing our effort to enlighten. This is the oft lamented confusion in economics between positive, i.e. descriptive, economics and welfare economics, which is pre-scriptive. To avoid the confusion it is better to ignore the theory of consumers' preferences and speak of preferences in the market ('aggregate preferences'). This confusion is an important impediment to progress. With its aid important economic factors, including defense budgets, regularly evade economic theory, and major factors such as religion and chauvinism and their influence on tastes and their arousing national and international conflicts are repeatedly played down. Defense and religion and similar 'external effects' were ignored in the hope that if tolerated they would go away by themselves.[6]

2. ERROR, SIN, AND OPTIMISM

The doctrine of natural goodness does permit 'evil' to occur – whether the evil people do regularly, evil which is for all to observe, or evil that superstition and war were considered to be. According to that theory no evil can be real evil, otherwise men would suffer from an inherent evil streak. Already Descartes, but more so Spinoza, accepted the Socratic theory of eudaimony, namely the view that people sin out of ignorance and not out of conscious intent. This was explained by the claim that 'sin' is harmful to the sinner. The logic of this explanation is faulty. No doctrine insists on the claim that sin is harmful to the sinner more emphatically than the traditional Western religions, yet these religions all claim that the sinner sins consciously and intentionally and maliciously, quite despite himself and quite against his own reason: the sinner knows that his sin will bring him eternal damnation, yet sin he will. Socrates, Descartes and Spinoza said no: when one knows how harmful an action is, one naturally avoids it; any other mode of conduct must be rooted in either ignorance or confusion, and confusion is a severe form of ignorance. They took seriously the claim that Man is essentially rational, that by definition Man is a Rational Animal.[7]

The optimism of this doctrine is obvious. It is indeed the optimism of the hope to have science and technology behind the new ideas, so that we can, indeed, view Descartes and Spinoza as disciples of Sir Francis Bacon, the optimist visionary of modern scientific-technological society. This may lead us again to the reductionist philosophy of Bacon. We can see the optimism of reductionism and the ability of its later advocates in economic theorizing, to bridge between people's current preferences or motives and their true ones. The more we understand and the more we can do – it was felt – the better the chance that we follow our true nature. The corrupt tastes and preferences we now have need not be opposed. Toleration is possible and commendable, since progress will leave old tasks behind.

Two central ideas, particularly important to those who wish to implement large-scale technological improvements, follow this line of thought. One follows classical economics and the other follows Karl Marx.

The idea which follows classical economics is that of foreign aid. It says the best foreign aid is economic aid. Interfering with other people's preferences is illiberal. It is also unnecessary on the assumption that helping them economically will bring them to improve their preferences directly. Marx had a similar view of religion. He assumed that religion will go away by itself with the improvement of material conditions and education.

Both ideas have been refuted: Local values may prevent the utilization of economic aid for economic growth, and economic growth may be accompanied by no or almost no growth of enlightenment or with the growth of an enlightenment different from that conceived by the philosophers of the Age of Reason and endorsed by Karl Marx, and paid lip service by his modern selfstyled disciples, Lenin and Stalin. The error of traditional rationalism which gave birth to these views is in the idea that social science should start as near the essential characteristics of Man as possible in the idea that Occam's razor should prevent the beard from growing at all. Viewing goodness as an essential characteristic of Man makes traditional-rationalism play down today's evil, and the power of evil in today's human affairs, on the doubtful assumption that when people are properly treated and given a chance to improve, they will improve. Moreover, if they fail to improve, the traditional rationalist tendency is to blame their ignorance. He finds consolation in the belief that when people are given an opportunity to learn they do so eagerly and successfully, and that then they necessarily improve. If they still fail to improve, then the traditional rationalist tendency is to blame their supersition as preventing their learning. And so, finally, in a desperate effort to make people worthy of our toleration, liberalism and favorable views of them, we come down hard on their religion and whatever else they deem sacred – quite nastily and quite unfairly. We do so with force and by cheap propaganda – thus qualifying as bungling politicians and as poor intellectuals.

The Enlightenment view of sin as error is deeply linked with reductionism, since humans by nature are described as good and as eager to learn, due to prejudice, and prejudice is due to the existence of established religion. Established religion is not part of human nature, though natural religion is. The claim that psychology is more basic than sociology erroneously led to the hope that the study of psychology instead of sociology would bring the solution. Since established religion is the enemy of reason, but only as a social institution, not as a personal affair, psychology is superior. The demand to avoid sociology altogether and appeal only to psychology is thus the demand to establish Man's Natural Goodness. Yet by the same token one should avoid psychology altogether too and appeal only to physiology. Except that no one quite knew how to do so, nor did anyone claim he could do so until Pavlov. This is why Pavlov became famous very rapidly though his salivating dogs were anything but a new phenomenon. His doctrine led to B. F. Skinner's utopian Walden II. We need not discuss the rights and wrongs of Pavlov. If he is right, then we should also try to do away with physiology and go straight to physics!

There is an answer to this too: we may practise sociology, but when psychology is available prefer it to sociology, and so on down the line. Thus, it was all right to practise associationist psychology until Pavlov showed that conditioned reflex is physiological, since we can train not only the mouth to salivate, but also the kidney to produce urine. To examine this answer, we need a concrete example which we can see in action. Here is a simple example which we have already met before. In his *The History of the Psychoanalytic Movement* Freud took pains to stress that he did not deny Adler's theory of people's wish to be accepted and the resultant inferiority complex they may have. But he felt this theory was sociological, whereas his theory of the sexual motive and the resultant Oedipus complex is psychological and so takes precedence. Now Freud's error is very serious and does not really pertain to whether we do or do not suffer from the Oedipus complex. Adler treated patients for their inferiority feelings and Freud for their Oedipal feelings. If Adler's treatment was superficial and successful, it is technologically preferable to Freud's deep analysis!

Freud would object. He said that we should treat the disease, not the symptom. Here tradition is on his side: the disease is always ignorance and treatment is always the acquisition of knowledge . . . in psychotherapy it is the acquisition of self-knowledge. Freud also had tradition on his side when he went further and said, ignorance is no disease, since we are all born ignorant. Disease is the impediment to learning, prejudice and superstitions. Not all prejudices, to be precise, but those which cling to the mind; this, he argued, is what neuroses really are. Here Freud also had the whole Hippocratic tradition on his side: medicine does not ever cure. Nature cures – Good Old Mother NATURE – and medicine at best only removes the obstacles on the way to cure. The idea that we must cure the illness not the symptom is very salutary because ever so often physicians are criminally neglectful by letting a malignant disease get out of hand while treating the pains it causes. This was particularly important for Freud in his early career, in his treatment of phobias. A phobia is a projection of deep-seated anxiety on some object which consequently seems terrifying. But showing a patient that an object is not terrifying will not do since the patient will then project his anxiety on another object: he needs his anxiety itself treated and that, of course, with the aid of self-knowledge.

Yet Freud was in profound error here too, and from the strictly technological viewpoint. Treating the symptoms of a malignant disease is poor medicine, no doubt, but treating the highly dangerous symptoms of a mild and passing disease is just right. There are many diseases which organisms overcome with ease unless they are killed by their symptoms: e.g. cholera and diphtheria and sunstroke. Freud took it for granted that treatment must go deep, and that psychoanalytic treatment must be complete: yet he admitted none of us to be perfect. On this point one could call him a fool were not the whole radicalist tradition of the Enlightenment behind him. Worse than that, he realized that the ideals of the Enlightenment were too high. And then, rather than lower his ideals, he sank into the depth of depression – exactly in accord with Adler's theory – while invoking the name of the great manic-depressive philosopher Arthur Schopenhauer. The inversion of the Enlightenment's high optimism into deep pessimism, with Mother Nature as revengeful, as divine nemesis, is today's hysterical flagellistic apocalyptic part of the ecological movement. This, then, is the chief technological corollary to the social philosophy of the Enlightenment. It saw physical technology as applied natural science, yet had no room for social technology as it saw social science as reducible to psychology. The reduction was both a matter of principle and of social technology. The Enlightenment Movement thus blocked all social technology on the assumption that it will prove unnecessary.

Example: Education is a complex system, which involves social institutions, psychology, and more. The social institutions in charge of education were almost exclusively religious. Unless one was a very aristocratic person educated by tutors professing secular convictions, one was at the mercy of religious educational institutions, receiving religious education of one sort or another. The American and the French Revolutions in which the principles of the Enlightenment Movement were supposed to be implemented created a secular, universal, compulsory education. The Enlightenment Movement condemned the established education system not as established, not as institutional, but as superstitious. To replace it the ideal of self-education was proposed. Bacon and Descartes stressed the evil of the kind of education they had received and pleaded to their reader to be educationally autonomous – to be an isolated intellect thinking alone, each for himself.

This idea is absurd, yet it led to the establishment of scientific societies which published scientific literature in a way accessible to novices, so that though universities in the eighteenth century were medieval, science flourished in the hands of amateurs. Yet the education system remained medieval. The Rousseauvian ideal of the self-educating individual impeded educational reforms. Rousseau himself added to the self-educating individual – Emile by name – a mentor. An adult individual could, indeed, join a learned society or a salon and find a mentor and even a benefactor if he needed financial support; but the child remained at the mercy of the educational system which remained terribly and unbelievably backwards. As L. Pearce Williams[8] observes in a classic paper, immediately after the French Revolution the French Assembly spent much time discussing education. It did so for two strong reasons:

First, the educational system which had been religious was taken abruptly from the hands of the religious authorities and badly needed a new administration.

Second, all the failures of the French Revolution could be blamed on popular prejudices, the fact that the people were not yet ready, etc., which called for education as the sole remedy.

Education, like so many other aspects of French life, was remodelled by Napoleon. He turned schools into military academies. We still suffer all over the civilized world from the Napoleonic reform of education. Of course his reform could not be that powerful were it not the major means of transforming the West from a pre-industrial to an industrial world: it made literacy the norm. Yet that is where we are stuck at present: the average Westerner is educated well enough to be able to read the instruction booklet and use the machine it describes. Today this is not good enough. We need a better theory of social reform and better social reforms, beginning with the reform of our educational system and ending with the reform of all our poor social institutions and organizations.

3. SOCIAL THEORY AND SOCIAL ACTION

Social science today is an outgrowth of many tendencies, most significant of which are two intellectual frameworks, liberalism and romanticism, an important political movement or two, nationalism and socialism, and, not least, the sheer necessity of coping with day-to-day problems. The fact that a few general ideas interplay in social science offers much leeway, but also much confusion. The liberals attempted to play down all social forces. The ideas of the state as a night watchman and of the market mechanism as the best regulator of the economy seemed paradigms of liberalism, as was the idea that religion and other 'superstitions' should be tolerated. These ideas, however, did not cover nearly enough. What else could the liberals offer?

Jeremy Bentham, disciple of Adam Smith, had the ambitious idea of a social science to satisfy all practical social needs. He claimed to be the Newton of the social sciences, with universal gravity being the law that we all need friends. Clearly, he was not the first to discover the universal need for company, but the first, says Elie Halevy in his monumental study of the growth of English radicalism, to argue with examples about the application of this idea to social reform, rather than to a revolution, as his radicalism might require. Bentham's idea is, indeed, still the cornerstone of civilized liberal legal reform: change the law to make it convenient and reasonable for any citizen, at least for most citizens, to be law-abiding. But Bentham's theory is very inadequate for ever so many reasons.

It is easy now, in retrospect, to survey the classical social studies, to see what contributions were made by members of what school of thought, and examine how much of such a contribution is acceptable to members of opposing schools and, if not, what substitution to that contribution they could make. This is particularly easy since the Romantics often simply reversed the judgment of the Enlightenment: they placed politics before psychology, the collective before the individual, and reduction from individual to society (for example, the psychology of a national type). Thus the trend is clear: the whole field of political sociology, of the role of government in society, sprang into action. Liberals could oppose, but only partly: they lost the initiative. Some liberals could replace repressive measures by voluntary association, but only with some ingenuity, and only in part. They hardly could do more justice to the nationalist movement – so much so, that many a thinker, such as Hans Kohn and Karl Popper still identify all nationalism with romanticism and see the growth of liberal nationalism as mere confusion. Yet Romanticism too did not fully encompass the rise of the nationalist movement. And because Romantic philosophers discussed nationality as given and as justifying certain rights of governments to govern rather than studying the growth of national movements and the process of nation-building and the accompanying social, cultural and economic phenomena, Romantic thinkers to date are at a loss to study the obvious strong interplay between technological situations and the growth of national movements. The same holds for socialism which has both liberal and collectivist wings, especially Western orthodox Marxism, now almost totally extinct. Marxism was partly romantic and partly liberal, as Robert C. Thucker has attempted to describe in his classic study of *Philosophy and Myth in Karl Marx*.

Politics aside, there were major upheavals in the Western world partly due to religious reforms in Protestant, but also in Catholic countries, including all levels and kinds of secularization, liberalization and backlashes as well, but mainly due to industrialization. Industrialization led to rapid economic growth, the redistribution of the nation's population and of the nation's wealth, and the development of new means of transportation. This way it contributed to the decline of what there was in Europe of the extended family as well as of many of the traditional social structures based on small-group interaction. This was accompanied by social unrest. The unrest drew the attention of social thinkers to the presence of alienation on a large scale, of new modes of exploitation with the aid of new cruel means, of the growth of new and exciting working-class cultures, and more.

Most of the social organizations we find today in the West and elsewhere is the outgrowth of social organizations that grew in the period of industrialization, either with no forethought at all or according to short-range plans, some to satisfy short-range economic needs of enterpreneurs as conceived and misconceived at the time, some to satisfy quickly certain explosive public demands. Most important aspects of today's social life, traffic jams in rush hours included, are outcomes of plans executed by managements and ownerships of industries, by local potentates and other personages, once most powerful and terrifying and now utterly forgotten, who imposed traditions pertaining to all important areas of life, namely, study, work and play. Some might wish to add religion to this list. One might say 'perhaps'. It is no doubt true that even such enormous social institutions as the Church of Rome, or even the Church of England, or smaller ones which are still enormous, often developed attitudes to central problems of social life as mere afterthoughts. Even such a terrible matter of principle as the Roman Catholic obscurantist attitude to birth control is not a matter of principle at all, but a mere historical accident or afterthought. The Church of England changed its attitude radically because Malthus had exerted an enormous influence in England; had his influence in Italy been comparable, things might have been different with the Church of Rome too. The Church that accepted Joan of Arc, Copernicus, and at a pinch even Darwin, could just as well accept Malthus.

The central fact about modern society is its modernity, and its modernity is conspicuous in physical and biological (including agricultural) technology much more than in social and political organization, its peculiarity in these latter fields is largely an unintended consequence, not a planned process. This is the heart of the matter. One can reach a few commonplace yet very significant conclusions from this. First, that Western society could permit the growth of science and technology. This is not true of all societies, yet today almost no society on earth will frankly bar science or technology as some science fiction societies do in the tradition of Samuel Butler's profound *Erewhon*. Second, that the growth of science and technology brings about social and political changes, at times drastic. Third, that these changes can be for the worse, arrest the growth of science and technology, or threaten with destruction. Whether the West or the civilized world, or the whole world for that matter, is reaching this stage is hard to say. When we can show for sure that some social factor is very dangerous we may look for ways to handle it. All sorts of accidents may happen, some of them threatening the survival of humanity or even of life on earth. Clearly then, since the problems have now become global, we may want to seek global solutions. In any case we may observe that the social structure of industrialized society is poorly planned because the prophets of industrialization have played down the social factor. We may therefore wish to re-design our society in order to make it more adequate for industrial practices. And the place to start is quite clearly from social philosophy; what we think a society at large is, how it might organize, what central social values it may have.

As a model we may take the pioneering study *Marriage and Morals* (1929) by Bertrand Russell. In 1940 this book was still deemed outrageous enough to disqualify its author from teaching in the City College of New York. In it Russell noted that the institution of the family was never designed. He did not discuss its origins and its changes through the ages, nor the variety of its manifestations all over the globe. Rather, he insisted, the value which is most strongly reflected in thinking about this institution in the West is that of St. Paul, which consists of two parts: first, that sex is undesirable and second, that since it cannot be eliminated it must be confined. The confinement, says Russell, is what is usually identified as the small family and the related sex taboos. Russell began by rejecting St. Paul's values. He made a highly revolutionary move by declaring sex neither desirable nor undesirable, but private – thus rejecting all views on the matter he could find in his society. And he declared sex subsequently to be constrained by the normal rules of moral conduct, neither less nor more. He concluded that sex as such has no confinement to the family. He asked then, should we abolish the family or does it have a significant role to play? We must abolish the family as we now (1929) know it, he said, and replace it with a new institution designed for the significant role(s) we may want it to play. This permits us to put on the agenda for the new institution of marriage both the roles played by the old institution and any new ones we may deem proper. The old roles may be modified as well. Russell glossed over this point and, ignoring sex at all except as a means of reproduction, he declared the role of the family to be the protection of the interests of children. This of course will not do for adults who want a permanent liaison late in life or who plan to have no children. Hence, Russell was in error. His contribution and his error may both be easily examined once we cease being scandalized by his libertine attitude toward sex, while rejecting it, as reject it we should, now that we know so much more about sex and reproduction than we did in 1929.

The re-design of specific institutions, however important, is barely sufficient for a society so open to so much gadgetry as modern Western society. For the gadgets offer new means of interaction between different social institutions and raise needs for new adjustments between them and for the creation of new ones. For example re-design of the family should adjust to the future and desired total equality of the sexes. What seems particularly important in this connection is to observe that a major corollary of the wild growth of social institutions during industrialization, each developing along its own lines in accord with its own local logic, led to the alienation so much bemoaned by the enemies of science and technology. Industrialization atomized institutions and thereby people: work in a factory was no membership in a community because the community was organized with no regard for the worker's needs. More precisely, his need to earn his bread was dissociated from all his other needs. He was invited to find in his factory no other satisfaction of any sort – nothing but his income; and thus he was pushed towards an extremely instrumental attitude to work. This seemingly accorded with the worst religious Jewish-Christian preaching about laziness and the reluctance of human beings to work unless forced by circumstances. Even in the later part of the twentieth century such stupidity as the religious Jewish-Christian views about laziness is endorsed, or at least taken seriously and used to defend such barbarisms as the maximal division of labor, the clocking of all workers in and out the factory at the opening and closing time, hierarchical and close supervision, and external quality controls, in addition to the waste of much of the potential of the female labor force by discrimination and segregation[9]. The absence of an enlightened social theory of work thus enabled people organizing industrial work to use the worst social theory of work when it suited them. The worst social philosophy separates work and play. It admits play only in the manner in which St. Paul admitted sex: reluctantly. And it separated both from learning, because learning is either sacred and so should be kept apart, or instrumental and so should count as the work of the apprentice. The idea that learning is sacred was endorsed by the Enlightenment Movement which declared it the ritual of natural religion and made popular Spinoza's description of it as the intellectual love of God. The reactionaries of the Romantic school thought learning suitable only for the elite, the ruling class, and for the genius who should rule. In the twentieth century attitudes to learning are very confused especially because learning is split into the arts and the sciences. The sciences identified with technology, and technology with high-prestige work if not also with technocracy.

We cannot make work, play and learning into one activity, since they are defined by their different ends. Yet we can see aspects of each of these activities in the other two. We can even move quite some way towards integrating them. Company and college football are such examples, and more so the cases of employees who take college courses at the expense of employers, not to mention educational games of all sorts. But the aim should be integration to a much larger extent, beginning with tearing down the walls of the colleges and the universities and making entry to them as free as technically possible.

4. THE SOCIAL FRAMEWORK OF SCIENCE AND TECHNOLOGY

Science and technology are in a sense quite universal, in a sense typically Western. The existence of societies almost devoid of technology challenges one half of this view. The Westernization of Japan and the existence of first-rate science in non-Western countries challenges the other half of this view. It may sound strange that Western technology is spectacular only since the industrial revolution – only since two hundred years at most! The idea so commonplace in the nineteenth century, that we owe it all to the steam engine is by now rejected. Today our view of possible energy sources for industry is so much broader since we think industry could have grown even if energy sources were small by comparison to steam engines, and had remained of the size of the windmill or the watermill. Marx's idea of the necessary concentration of capital due to the inescapable economies to size is an oversimplification. Even the very importance of science for technology has been challenged by the claim that so many technologically significant ideas were not rooted in science at all, such as the use of vulcanized rubber for so many purposes, and almost all agricultural innovations up to the last few decades except chemical fertilization. The same held for almost all medicine until the discovery of vitamins less than a century ago – in so far as until then medicine was any good at all. This can also be said of the invention of the modern road and railway, of the automatic loom and myriads of similar automata, of nineteenth century foundries, and even of the invention of photography and of the phonograph. Nevertheless the scientific character of modern technology is clear and the ethos of technology which seems to have been a very important impetus was also created by Francis Bacon as part of the scientific ethos.

What social framework made it possible for this philosophy to become a movement is a fascinating historical question, deeply linked to the question, what should we retain if we wish to retain the scientific and technological character of Western Society? Not long after the establishment of civil government in the newly-founded regime in Soviet Russia, in the inter-war period, the idea of planned science was born. This idea attracted many Western scientists who consequently became supporters of Soviet Russia and of its official philosophy. In opposition to this a physical chemist, Michael Polanyi, was instrumental in the development of the Committee for Cultural Freedom and the Committee for Scientific Freedom, where the freedom in the title meant the opposite of planned science. Polanyi himself relinquished his position as a professor of physical chemistry to become a sociology professor. Since then and till the end of his intellectual career Polanyi wrote about the need which the scientific community has for intellectual autonomy. To this end Polanyi changed his career yet again to become a philosopher of science.

Polanyi's commitment to democracy is unquestionable, yet this commitment to intellectual freedom may be questioned, and it may even be asserted that his opposition to planned science was more an opposition to communism than to planned science proper. His contention against planned science was that only scientists, not politicians, can make reasonable assessments of research projects and make research proposals. This is hardly a criticism, since no one ever dreamed that the Russian political leadership had the ability to assess research projects. Everyone interested could know that the Russian political leadership was not aware of the first thing about the natural science of its day, so that it could not comment on research activities there, let alone voice comments judiciously. Nor was this ignorance peculiar to the Russian political leadership or to the period in question. Lord Snow reports he was amazed to hear that soon after the war British scientists secured permission to explode atomic devices from a prime minister not even aware of the existence of radiation fallout. His amazement was not at the ignorance of the then British prime minister, but at the degree of the deception leading British scientists could and did exercise in their irresponsible lust for power and prestige.

Planned science, then, is planned not by politicians, but by the leadership of science, namely by the leadership which guides the rank-and-file scientists in their research. This is well in accord with Polanyi's own view of how science is conducted anyway, a view now broadly received in the West as the idea of Thomas S. Kuhn. Kuhn calls the rank-and-file scientists normal and their research puzzle-solving, as opposed to the leadership of the scientific community which he calls the scientific leadership and its task which he calls paradigm forming or paradigm change, or scientific revolution. It is like Bambi's father: it appears at a time of stress and disappears when things look like getting back to normal. To call a research worker with a slave mentality normal is, unfortunately, to prescribe the norm; to call the leadership scientific is likewise meant to sanctify leaders – even leaders who are ignorant – and transmute them into learned people ex officio. The view Polanyi and Kuhn represent is clearly establishment. The only difference which demarcates the advocates of Russian-style planned science from Polanyi, Kuhn and their Western adherents, lies in the source of the leader's power. Russian-style planned science forces scientists to receive their authority from the Party, whereas in the Polanyi style version the scientific community is politically, but not economically, relatively autonomous. Now the Russian-style scientists are fairly autonomous too if they excel, as physicists in the better days of Landau, or are dominated by lackeys of the party, as in the days of Lysenko in biology, and usually something in between. But in the West too, science is autonomous, but it is not utterly free of external influences – chiefly party-political – as Polanyi himself bemoaned. Moreover, with both Western and so-called 'socialist' countries becoming increasingly technocratically oriented, this difference is bound to diminish (James Burnham, *The Managerial Revolution*, 1941). Suppose it was diminished. Would Polanyi then object to planned science? Presumably not, since all his criticism would then be taken care of. This might suggest, then, that his democracy is quite external to his view of the autonomy of the scientific community. Let us examine this.

What is the structure of the scientific community? How is this related to technology? This is a sociological question, and much hinges on what one means by community, since undeniably scientists are a social group of sorts. Come to think of it, it even depends on what is meant by 'scientist'. We may admit the existence of a scientific community and declare all true scientists not members of it. The scientific community in Nazi Germany, for example, is generally viewed this way; all real scientists there and then were either in exile or in isolation. It is usually suggested that the scientific community was not really scientific. Einstein thought otherwise and saw all German scientists who stayed in Germany, except for a handful, as culprits. The popular view, which goes the other way, nevertheless shows that one might assume the existence of a scientific community and claim all good scientists are excluded from it.

What makes it hard to view all good scientists as outside the scientific community is the realization that scientific activity is more complex than involving research alone, that at the very least it involves scientific education. Polanyi stresses this aspect or task of the scientific community. Yet it need not be so. In the Age of Reason scientists were educated, if at all, in the anti-scientific universities; they picked up their scientific education as amateurs, both from the literature and from scientific gatherings of amateurs in scientific societies. Universities offered, at best, courses in most elementary mathematics, astronomy, biology, and medicine. Except for Lichtenberg and Black no one used experiment in the class-room before 1800, and the first research laboratory in an English university was the

Cavendish, founded in Cambridge in 1871, which, Maxwell observed, was very poor indeed. Professional science is the product of the secular universities of the nineteenth century. Professional science in industry is contemporary with it. Professional scientific or technological research in a military establishment is only a recent phenomenon. The National Science Foundation, the Atomic Energy Commission, the Space Program and their European counterparts such as CERN, which employ enormous batteries of scientists and scientifically trained technologists, have changed the character of the scientific profession and have changed our image of science and technology, of the relation between them and of their social setting and their social and political import. They are all so young that their absence is still a living memory of colleagues who painfully and hopelessly looked for intellectually-oriented jobs in the thirties. These institutions are nowadays the mighty center of what Kuhn calls normal science.[13]

The autonomy of the scientific community which Polanyi and Kuhn discuss is the autonomy of decisions concerning what research to conduct. Who decides? How can a government or an industry spend millions on research and still not interfere? Answer: only by special arrangements, instituted by political or civil contract. Yet many governments and corporations decided to support technological research while bribing universities to pressure professors to cooperate. Polanyi advocates that researchers stay reasonably financially independent and reasonably free to associate in their internal clubs; and no one can object. Yet he approves of pressure to conform and Kuhn elaborates by spelling it out: the pressure is the threat of loss of a job. Job means income, and income is usually financial dependence, not something given to the total discretion of the leadership to manipulate and use in order to exert pressure. What happens is that administrators put pressure on researchers. What Polanyi and Kuhn hope to see is the universities being enormous financial concerns, employing hoards of scientists in research and in education, and exacting from members of their communities enormous efforts by the threat of firing them, while allowing for no pressure from society at large, and no control from the community over research. The social philosophy of science advocated by Michael Polanyi and Thomas S. Kuhn is not democratic, does not attend to the needs of society and allows no one but the scientific leadership to attend to them. The scientific leadership is the university and research institute administrators, most of whom happen to be these days defunct, disenchanted or failed researchers. In brief, we have here a veiled theory of research organizations as university-based and of the university as an autonomous medieval guild.

The popularity this theory enjoys among 'normal' scientists reflects not the independence of the university and its research activities; rather it reflects the university as the power base of the new scientifically trained political go-getter – the new technocrat, the person trained to be a researcher who prefers to do something else. By contrast, the medieval and even nineteenth-century college administrators were part-time or temporary functionaries; they did not give up their little teaching and their less studies.

5. CONCLUDING REMARKS

The philosophy of the Enlightenment was liberal and individualistic and therefore very appealing. Yet its appealing traits were not backed by reasonable social scientific and technological ideas. Today even the simplest economic planning is impeded by monetarist economic theory whose appeal is derived from the philosophy of the Enlightenment Movement, especially in the United States, whose Declaration of Independence expresses the principles of Enlightenment. Yet modern society calls for planning, not only of its economy, but of all its social and political institutions, so as to make technological society flexible and humane.

CHAPTER 10

POLITICAL PHILOSOPHY AND ITS IMPLICATIONS FOR TECHNOLOGY

The most obvious fact of international political life is that the highly technologically advanced nations, the industrialized nations, are parliamentary democracies and science based, that the communist states make serious efforts to modernize – to advance technologically – and overtake the highly industrialized parliamentary democracies, and that on the whole the international political situation is such that few dare take stock of it at all.

Is democracy essential for technological progress or industrialization? Is it inherent in rapid successful technological growth – industrialization – that it be capitalistic? Most people say, yes, though, under the influence of Lenin, Marxists have modified their affirmative answer. They now say socialism may precede industrialization and then accelerate industrialization even more than capitalist parliamentary democracy – like a child skipping a grade.

The Third World, so called, the poor backward suffering half-starved majority of mankind, suffer either capitalism or communism as an intrusion. Quite clearly, if a philosophy of technology has a poltical component of any significance, it should be its view of the politics of technological transformation. This problem is not only of the poor countries; it is, to use Buckminster Fuller's excellent expression, the problem of maintaining Spaceship Earth.

What has political philosophy to say to those concerned with the use of spreading technological advances for the relief of urgent global problems? What is the proper philosophy for technology transfer?

1. THE JUSTIFICATION OF POLITICS IN ACTION

Traditional political philosophy still engages most political philosophers and many political scientists. Yet it never was very practical since it asked the question, What legitimizes political power? If we cannot answer this question, anarchism seems the right political theory: no government is justifiable as long as it uses force; therefore the only government permissible is one which uses no force at all. The word "anarchy" is misunderstood to mean no order. This is an error. The word means no rule. It is the anarchist's view that we can have order and a proper administration run voluntarily, namely, without the use of police and army. Hence, anarchism is either pacifist or in favor of keeping a voluntary national militia for defence against foreign invasion until the whole world has gone anarchist.

The implicit assumption of most classical political philosophy is anarchist or near-anarchist: no justification of governmental force is possible except to prevent the unjustified use for force. This presents the celebrated theory of the state as a night watchman or a policeman as a version of near-anarchism. This theory gained the authority of John Locke. Its leading advocate is Adam Smith, the great political economist who firmly believed that the greatest amount of wealth will accumulate if the government will do nothing except enforce peace and the keeping of contracts, so that trade and industry may flourish. The answer to Smith's view is simple: vast areas of the earth saw almost no government to speak of, yet capital did not grow there and trade and industry there are minimal. Smith knew all this, of course, and required not only that government be minimal but also that prejudice be removed and science and technology accepted as ways of life. This may very well be true, yet it does not in the least help the poor people who occupy vast parts of the earth. On the contrary, the ideal accepted today in so many countries is that of Dr. Sun Yat-sen, the father of both Nationalist China and Communist China. He wanted from the West only guns, not culture, not science, nor democracy. (He experimented with democracy and thought it a failure.) If Adam Smith is right, then the question of industrialization is transformed into this: How can we convince others to become scientifically minded? His view that the scientific attitude leads to a free market is also an error: communist countries, including China, are scientifically minded but anti-capitalistic. Whichever way we look at it, Adam Smith's idea, whether true or false, is not useful.

So much for the idea that a government's force cannot be legitimized and for the consequent near-anarchist theory of the state.

The idea that some government force can be legitimized brings the hope of distinguishing between good and bad government, and with it the hope of finding a criterion for good government. The criterion may prove dangerous in that it may justify only a utopia. In particular, if the justification rests on rationality – on scientific rationality, that is – then inescapably political philosophy will justify only scientific utopias. And, of course, as long as scientific rationality was deemed compulsory and commanded unanimity, there was no escape from what Jacob Talmon has called totalitarian democracy – namely, from the populism of a science-dominated society. To be more specific, scientific rationality permits disagreement, but only on questions to which science has not offered decisive answers. How could parliamentary democracy grow under such conditions?

The answer is complex. The rationalists did not recognize the nationalist movement; they did not pay attention to the actual processes of government which were seldom so tyrannical as to allow for no dissent, groupings of dissenters, power struggles of interest groups, and the like. Also, much of Western political philosophy is an afterthought to the views of a few great thinkers who did not limit their ideas to the traditional problem of the rational legitimation of power. The greatest among them were Macchiavelli, Hobbes, Spinoza, Locke, and Montesquieu, as well as de Tocqueville and Mill.

Macchiavelli was a true son of the Renaissance. His aim was to return to the old glory of the Roman Empire; he thought that the Pax Romana justified this well enough. He had no intention of acting politically. Rather, he sought to be a guide for a person motivated enough to act on his own accord, and intelligent enough to take advice. The idea was truly Macchiavellian. Macchiavelli's advice had to perform two tasks at one and the same time – the advice had to appeal to the ambitious power-hungry leader and help him advance, and the advice had to steer the same person to further the interest of the author in addition to his own interest. The general idea was to help a person develop enough political power to unite Italy, control it, and maintain the peace. Moreover, it would be in his interest to control his immediate subordinates by fear but the population at large by love. And this would be Macchiavelli's way to establish the once-and-future Pax Romana.

Like Macchiavelli, Thomas Hobbes wrote under the stress of a civil war. He justified a dictatorship as long as the dictator maintains the peace, thus cutting short Macchiavelli's deliberations and coming quickly to similar conclusions. The citizen's private affairs, advised Hobbes, should be presumed innocent and of no interest to the government.

The next step is complicated, since it was taken by Baruch Spinoza, whose influence cannot be clearly assessed. He was a Jew, a heretic, and a political radical; clearly a dangerous fellow to associate with. (More than a century after his death, Kant found it necessary to repudiate the charge that he was a Spinozist; there is a handsome complement, at the end of Kant's major book, his *Critique of Pure Reason*, which is clearly given to Spinoza even though his name is not mentioned there at all.) Spinoza recommended liberalism, toleration, parliamentary government – even a constitution in the modern sense of the word though without using the word, not yet coined. (He also recommended high taxation, regulated by freedom of migration. This free enterprise argument has not been noticed by any capitalist theorist, from Smith to Milton Friedman.) Although it is not clear how much, it is clear Spinoza had some influence.

John Locke was a philosopher within the traditional mode, the originator and defender of the watchman conception of government. It is well known that his idea and idiom influenced the Declaration of Independence of the United States. Yet, following the ideas of Macchiavelli and Hobbes, he made central the idea of the separation of powers, as a means of control over arbitrary governments. This idea is more practical than Spinoza's.

Modern democracy is inconceivable without Locke's checks and balances – based on Macchiavelli and later developed by Montesquieu. Separation of powers, not fully practicable, is practicable enough to guarantee the possibility of checks and balances. There has not been enough study for us to know how checks and balances are supposed to be integrated into the watchman state: a watchman only watches and needs no power to do anything else, no checks (except that he does watch) and no balances. But checks and balances can work, whereas the watchman state is a remote utopian ideal. Contrary to Marx's claim that Smith was the spokesman and the herald of the capitalist state, the capitalist state proper – the utopia of utterly free trade described by Smith – never existed. (Smith did declare that the Hanseatic League was a model of sorts for his theory.) It is only because the ideal liberal state never existed that de Tocqueville and Mill, spokesmen for liberalism and particularly for minority rights at the very peak of the allegedly capitalist era, were so important.

It is necessary to mention here the traditional irrationalist, anti-liberal, anti-individualistic, authoritarian, holistic political philosophy of the Romantic school. That school contributed holism to science – both physical and social – but it also made individualists, liberals, and rationalists all too ready to reject holism because of its origins.

So much for traditional political philosophy. What it tells us about technology is only one thing: the capitalist marketplace creates incentives for technological innovation. This, as anyone who has examined the economic problems of inventors should know, is utterly false; there is no free entry to the market even for inventors. Unless one invents the transistor, the laser, or the splicing of genes, nowadays one must sell one's invention to big corporations or leave it – one cannot enter the market today with a reasonably worthwhile innovation and strike it rich unless one is backed by big capital.

(Recently economists have noticed that Smith's model does not even allow for innovations, but this is a technicality that need not concern us here.)

Popular literature for over a century has attempted to connect technology and politics, and the advent of the nuclear age fortified its two models. The more popular model concerns a powerful invention, used by its inventor or by others who usurp it, to take over a country from its legitimate government. This plot in fiction is no more than a variant on the true story of the Guy Fawkes rebellion. The only difference between Guy Fawkes and Jules Verne's Captain Nemo or H. G. Wells's Invisible Man or Dr. Moreau is that the real character did not act alone but as a part of a large group. In other words, the pattern is that of a conspiracy theory run wild. It has little or nothing to do with technology; indeed, transplanted to a setting in the Middle Ages, the plot proceeds well with no technology to speak of. The other model is that of a technologist, an engineer or a physician, who uses his skill for social climbing. This gets slightly more interesting when, in order to further his social ambition, the technologist creates his own organization – a modern cigarette factory, a hospital, a new government agency. Here too, technology is but a pretext; the story's interest lies elsewhere.

Classical political philosophy, then, is as irrelevant to technology as it is to other aspects of modern life. Notwithstanding its popularity among philosophers, political thinkers, monetarists of the Chicago school, and others, it has made no positive contribution to the study of technological growth.

Perhaps there is an exception in tales about how science may take over government (usually world government) so as to establish a utopia (usually a non-democratic minimal government). Interesting versions of this story are Isaac Asimov's *I, Robot* – in which he questions the moral worth of a decision making procedure relegated to computers – and Bertrand Russell's *Nightmares*, where American and Russian technologists conspire and coordinate their efforts to keep the arms race going on forever. Most writers of science fiction, indeed, have advocated government by science. If this has any relevance to real life, it shows how beneficial the extensive use of physical technology might be, or how useful it would be to turn government over to scientists – so that they could create a technological utopia at once.

2. THE POLITICAL INEPTITUDE OF CONTEMPORARY LIBERALS

The problem of all liberal philosophy is that of the rational legitimation of authority simply because liberal social philosophy declares every person his own authority; it denies, *a priori*, the authority of any one individual over another. In real life, this is not the case. In real life, heads of families still terrorize their charges, and the beating of wives and of children is still a common phenomenon even in the most civilized sectors of the most advanced liberal societies. In real life, employers and managers have enormous power over their employees, whether sanctified by law and custom, or opposed by law but retained by custom, or merely backed by the logic of circumstances. And, in real life, the law exercises both just and unjust authority over the individual citizen in myriad ways. From the very start – as Sir Leslie Stephen has observed in his *History of English Thought in the 18th Century* – the utopian bias of traditional rationalist political philosophy, its effort to see the world through the eyes of the perfectly rational man, has led to systematic ignoring of women, children, servants, peasants, employees, and the like. Political philosophers asked why do I and people like me accept the authority of the current, real, all too stupidly run, political institutions? To this question they had no cogent answer because they were utopian. Yet the question calls for real answers concerning real situations of real people. Confined to idealized, rational subjects, the question led to answers in terms of idealized rational government. This, said Spinoza, made all political thought irrelevant to real politics. Yet idealized political thought is, by now, part and parcel of the modern political world. Just as, because of their enormous popularity, we cannot brush aside Catholic or Marxist political philosophy, so also we cannot brush aside classical rationalist idealized political philosophy. All are important parts of the vast tapestry that is the contemporary political, social, and economic scene.

Moreover, though Spinoza rejected ideal government, he did endorse the ideal of rational man; he even noticed all too clearly that such a man needs no government at all. He said laws exist for the education of people who are not yet rational. Since the ideal rational man is a philosopher and a scientist, the community of scientists needs no institutions of either government or control. Ideal liberal images of the political system follow the ideal of rational man; hence the political system must be more or less anarchistic.

The following simple consideration illustrates this. Classical economics suggested that controls over the economy are redundant. The market mechanism is the best means of control of the economy as a whole. The consumer can be trusted to opt for the best deal available; he will not purchase inferior goods when better substitutes are available at the same price or equally good substitutes for a lower price. This has led some modern followers of the classical school, the neo-classical or Chicago school of economics, to oppose government quality control of products, truth in advertising, and similar government services to the consumer. However, these neo-classical economists do recognize the government's duty to insist on the execution of obligations undertaken by contract. In other words, when a customer agrees to purchase inferior products, the government is not required – and therefore not allowed – to interfere; but if he contracts to purchase a quality product, and receives an inferior product instead, then the government may be required, may be obliged, to interfere. This reasoning is inconsistent since the misleading labelling of a product may be taken as a part of the contract of purchase; and the public, then, may demand truth in advertising. If so, then the government is definitely invited by the public to exercise quality control. But, and here we must agree with the extremist members of the Chicago school, this is the thin edge of the wedge. Once, on such a liberal interpretation, the government is invited to interfere, there is no end in sight. One thing may lead to another if the public opts for all sorts of controls. And so some extremist members of the Chicago school accept the consequences; they say we need no government intervention at all, not even in the prevention of the violation of contracts. People who violate contracts will simply be known as liars and others will avoid doing business with them. Other extremists of the Chicago school say this picture is too rosy; in a large modern market we have to be able to force people to stand by their contracted obligations. But, they add, we do not need a government to impose this: we may use the market mechanism. This means that we can hire policemen in the market to force our business associates to pay us their debts. (This amounts to a demand to legalize the loan shark's practice of hiring thugs to use violence against customers who cannot pay.) But, then, how does the policeman know that when he does his job he will get paid? The Chicago school is forced into this position because it endorses the theory that there is enough rationality in the market to restrict government action, and then tries to apply it to real markets that are, as we know them, wild and bizarre.[10] Hence we can say that classical liberal political philosophy is more or less anarchistic depending on the degree to which members of society deviate from the ideal of rationality – all based on the grand, and palpably erroneous, assumption that the deviation is marginal.

Looked at another way, we can well imagine a society comprised almost entirely of reasonable, articulate people, glad to debate matters and settle disputes amicably, and thus almost in no need of police. Such an image is offered in the science fiction novel *The Black Cloud* written by the famous astronomer Fred Hoyle. This novel is an important document of our time. It is immensely popular among scientists and technologists in research and in industry; it is deemed a primer of the social ethics of science; and it is frankly and thoroughly contemptuous of the broadly accepted institutions of Western parliamentary democracy. Elitism and democracy, indeed, do not mix; and technocracy is but a form of elitism.

What does a piece of fantasy matter? After all, it is the case that scientists and technologists are, at least in democratic societies, paragons of law-abiding citizens, often conscientious to a fault. For example, one reason that the scientist J. Robert Oppenheimer incurred President Harry S. Truman's contempt was that each of them insisted on taking full responsibility for the Hiroshima and Nagasaki nuclear destruction. Oppenheimer refused to recognize that responsibility clearly lay in the hands of the President; he insisted on feeling personally guilty.

The guilt Oppenheimer felt was outside the rationalist tradition; its introduction is a violation of the rules of the game. The perfectly rational have no sense of guilt and there is no room for it in their emotional makeup. The violation of the rules of the game by the introduction of a sense of guilt, however, is not the beginning of a new system. Rather, it is the introduction of a small sub-game, of an epicycle on a cycle, meant to save the old system. Once we realize that the grand masters of Los Alamos and the Manhattan projects were sinful and felt guilty, we may view their case as a minor deviation from the norm, a correction of it – both understandable in the pre-history of true technocracy.

So we need to correct the picture and replace Robert Oppenheimer, the almost ideal rational man and technocrat, with Fred Hoyle's piece of fiction. His daydream rectifies the faults of true history. Hoyle's scientist acts rationally and is not deterred by all the politicians of the world; he is not willing even to bow to the majority of his own peers when he knows that he is right. He is able, responsible, decisive.

The popularity among scientists and technologists of Fred Hoyle's daydream, and the popularity among them of the daydream that his daydream is reality, are the excuses they make for their failure to face responsibilities. The normal scientist abides by the laws of his land and lives by the customs of his community. He does not quite belong, but he does not dare violate the laws – unless he defects from one country to another, which makes no difference at all. He makes his living by the fruits of his research. When he feels saintly he worries about the possible ill effects of his work; then he feels an inflated sense of responsibility, as if he is in charge, and then he feels guilt. He then regains his sense of reality and continues as before.

Exceptions to this grim picture are the heroic Russian intellectuals who preach democracy within a setting that makes this activity a cause for persecution and terror. It is pathetic that in democratic countries sentiments for democracy are aroused in the scientific and technological community almost only when rallies are organized for the sake of these Russian heroes.

Why is this so? Why are Western liberal leaders of science and technology so inept when it comes to safeguarding and improving democracy in their own communities and countries? Why, as Einstein and Russell wondered, can they not be organized to improve the chances of world peace and survival?

At least a part of the answer must lie in the fact that current ideology in science and technology is confused; it vacillates between classical radical liberalism ("right radicalism," so called) and classical Marxist technocratic socialism (left radicalism).

(These two philosophies, incidentally, are equally devoid of ideas concerning international relations, including even such simple matters as international trade – especially of fuel – and foreign aid, not to mention matters of war and peace, environmental and population control.)[10]

Consequently, most leaders in science and technology are, as often as not, apolitical, even anti-political. Their political activities run along two rational yet conflicting lines: they try to minimize the government's control over citizens, and they try to increase its efficiency by the use of modern high-powered technology and expertise. The second line leads to technocracy; the first only limits the applicability of technocracy in cases in which conflict arises between individual citizens and the state. This is a policy of having no political initiative.

"Science and technology," as presented here, is the pastiche of popular prejudice. It includes pure natural science and applied natural science and technology – heavy on hardware and public administration but with no social science proper and no political science at all. So enhancing this prejudice only enhances the status quo, the political impotence of scientists.

The current dogmatic adherence to traditional liberalism thus renders liberal scientists and technologists politically impotent. Some of them, especially in the United States, endorse one utopian ideal or another along-side their liberalism – usually socialist. The relevations about the Gulag have made them less friendly to Communist regimes, yet they remain utopian. Their utopianism enhances their political impotence since, by comparison to the ideal, the present and the slightly better future which we might work for are hardly different from each other. To them the game seems not worth the candle. Others go further and explicitly forbid the improvement of the current situation as conducive to the enhancement of the present negative state (Herbert Marcuse). In particular, every effort to improve the quality of working life is condemned; it invites the cooperation of workers and management and the enhancement of whatever interests the two antagonists share. In days when labor accepts cuts in remuneration in order to prevent the closing down of plants, not to fight for the humanization and the democratization of the workplace is sheer political impotence – and any theory that supports this is an illusion.

So it becomes increasingly hard to say whether theory – classical liberal, socialist utopian, or a combination of the two – is conducive to political impotence or is a mere cover-up for it. Certainly many technologists are just not interested in politics with no excuse at all – or with the flimsy excuse that technology, being means, not ends, is morally neutral and capable of being used by good or bad parties to good and bad ends.

In truth, ideology is neither the cause of impotence nor an excuse for it. Ideology is usually inadequate for action and the very endorsement of it is impotence as such; otherwise it is adequate and potent, is power, indeed – political power itself. On this Karl Marx was right, though he obscured it by his metaphysical wording: ideas materialize when accepted by the masses, he said.

We need new solutions to urgent problems, and for this we need a new and more adequate intellectual framework. Since the chief political end of every society is survival, and since the secondary political end of every society is its welfare, peace, and the maintenance of its values, developing a new framework should begin by discussing these ends and what they demand of technology. Many a society was destroyed – out of an inability to control its environment, its inner structure, or its technology. Technology is supposed to help a society control its environment by offering it tools. This supposition makes modern technology the production of hardware, so that modern technology may weaken a society's inner structure. This is Marx's view of all political revolutions. But, contrary to Marx, the inability to control technology has often led to the total collapse of societies. And, contrary to Marx, social technology may bring about, peacefully, some essential improvement. This should be some encouragement to those worried about the possible destruction of the eco-system as a whole – whether by pollution or by nuclear war.

The new framework, then, should include ideas of technology to control current dangerous technologies. To this end we should examine current views on how current technology should be controlled. It is a fact that technology as a whole is not controlled – is out of hand; and that political philosophy should tell us to stop developing technological power as a whole, to begin to develop means of controlling it as a whole. How and why was the control of technology lost? It was never lost; we never had it. Rather, our present question is, what suggestions are there to reduce the effort put into technological growth and increase the efforts put into the growth of the control over technology as a whole?

Many say it cannot be done. They insist that the machine has now become more powerful than mankind and is threatening to destroy all. This is a metaphor and a very dangerous one. Machines have no goals of their own, and the only machine we have lost control of is us. (We have found the enemy, says Pogo, and he is us.) The answer given by Aldous Huxley in his *Point Counter Point* of 1930 and elsewhere is essentially a moralistic one: we should moralize politics, since the trouble is that politicians do not care for humanity, only for their private immediate interests, and they hoodwink the public. The obvious moral he drew from this was that we should go to the public. The result was disappointing. Rachel Carson, the pioneer ecologist (*Silent Spring*, 1962), was first ignored and then admired, yet the public is not moved to action. Why? Indifference. The public, like politicians, are shortsighted. Moral: explain to the public. No result. Why? A mixture of a lack of imagination and a lack of concern. Moral: dramatize. Dramatize even more. Describe detailed horrid scenarios.

Still no result.

What is wrong here is the absence of a political approach. Even were the public pressed into action, no action is suggested. The energy conservation program has failed and was doomed to fail. The only way to make people save energy is to raise its price. Governments find this impossible to do internally. When energy exporters raise the price of energy, the result is inflation, not saving. No other program has been offered. The objections to all of them are rather obvious.

The truth is, as Jacques Ellul has found out, that we are all lost. The danger is rooted in our reluctance to admit our inability to cope, our inability to devise a program that has any chance to work. All existing plans are stop-gap, but their proponents pretend that they are proper plans. This predilection must stop if we seek change. This is the first assignment on the way to change. Ellul says that it is vanity to pretend that the technological process can be checked or guided. But it is equal vanity, it should be added, to pretend that the process cannot be guided. Confession of ignorance is in order, and the search for a possible way out may thus begin.

3. POLITICAL THEORY AND POLITICAL ACTION

Political philosophy is the traditional study of the justification of government and its limits. Political science, for the most part, classifies different systems of government, of political machineries such as political parties, and of administrative techniques. Meanwhile political action is a spectacular game of roulette, often played by ambitious egomaniacs. All around the world military rebellions are repeatedly effected and are given the honorific name of revolutions, especially when accompanied with the selling of one's country to a new buyer. The question is, where is the roulette game and who plays it? For example, when it turned out that the national broadcasting service or the governor's palace was the political roulette wheel, the seat of power, colonels could not be expected to resist the temptation to take over the seat of power. When counter measures to such takeovers were discovered, it turned out that rebels could use similar counter measures too. Why is this instability not examined and contrasted with the stability of industrialized countries so that the unstable countries can stop these games?

Because the theory is a mess.

According to Karl Marx, the rich owners of industry rule the world, since they can purchase guns. And politicians can then either be bought or shot at. This is not the way Marx put it; rather it is Bertrand Russell's translation of Marx's ideas into his own language. Once so translated, Russell had a simple response to it. It is not that the rich buy guns but that gunmen rob the rich; not the rich industrialists but the powerful generals rule the world. The American sociologist C. Wright Mills was an admirer of both Marx and

Russell. He was here in a dilemma, since he did not want to side with one against the other. So he found a solution: he said that the rich owners of industry and the strong owners of guns are not different groups, one of which should control the other; rather the rich and the strong are united in one small indivisible interrelated intermarried social stratum. C. Wright Mills called the stratum by the now famous title, the military-industrial complex.

But the major confusion this fashion is meant to create is between observation and theory. Power is observed to be distributed, and some theoreticians have tried to find its source – claiming, for instance, that the source is much more limited than the observed broad distribution. What is the presumed source of political power?

Traditional liberal political philosophy, in its study of the legitimacy of government, locates the source in each individual. By right, every individual may govern himself only, and his transfer of this right legitimates government. The Romantic school challenged this as too individualistic and based the right to govern on the ability to govern, thus putting the theoretical source and legitimacy of power in the hands of those observed to be powerful. This is *Realpolitik*: might makes right.

Marx had a better idea. He was not concerned with power elites; he did not care who wields power. He claimed that, when in conflict, economic interests overrule other interests. If a person in charge of an economic interest is not power-hungry, he will not wield power and not exhibit supremacy. But when some other interest threatens the interest he is in charge of, he will be forced to move, said Marx ("On the Question of Free Trade"); he will act to constrain the opposing interest.

(Hence, when Bertrand Russell translated Marx's view into his language, perhaps something got lost: the claim of Marx was not a matter of power, much less of the question in whose hands it resides, but of interest – whose interest has priority? Alternatively, it was simply Russell's confusion of Marx's view with his own: the supreme interest, said Russell, is not economic but the need to control – power.)

Marxians say that no action is useful unless it advances the socialist revolution. Hence, this should be the end of demonstrations organized in capitalist countries by people who have the interest of the environment at heart. The capitalist economic interest overrides the interest of the ecology and it will win. In what way it will win is a secondary question. People may force capitalist governments to cancel licenses and plans for nuclear power plants. But then a scarcity of oil occurs – perhaps real, perhaps artificially created by the oil companies, perhaps artificially created by the government, perhaps artificially created by the oil companies at the instigation and/or blessing of the government. It does not matter. When the population suffers from a scarcity of oil, it will be time to reactivate all sorts of plans shelved under pressure from environmentalists who had organized mass demonstrations. Perhaps things operate quite differently. But as long as the economic interest takes precedence, Marxians say, the ecologists' protest is at best temporarily effective. All this is convincing enough, though it is not in the least related to capitalism or to socialism.

Someone might object that the interest of rescuing the environment, after all, is an economic interest too. Karl Marx answered this question: the economic interest which capital takes care of, he said, is immediate. As long as there is conflict between long-term and short-term interests, the long-term interest will lose; the system will collapse. In socialism, however, short-term interests and long-term interests coincide, Marx guaranteed.

Anyone who wishes to have effective influence on political life must indeed take heed of this admittedly rather naive part of Marx's theory. The weak aspect of it is that it says that priorities cannot be tampered with: the capitalist is in charge of the economic interest, and the priority of this interest is neither questionable nor alterable. Yet something is alterable – even according to Marx. Interests, he said, are class interests. You cannot change interests. You cannot re-order their priorities . But you can change the ruling classes, and it is the ruling class which decides what interest has priority.

Marx was in error. There are class interests, as he said, but there is also the national interest, contrary to his teaching, and even the interest of mankind. Also, there are vested interests of all sorts of social groups and strata, including that of the power elite, which Marx deemed always and in principle powerless since only classes have real interests and only real interests wield real power.

Yet there is wisdom even in this serious error of Marx. Although very often direct action is called for, at times indirect action is wiser. Some legislation alters people's modes of conduct. Some economic changes alter some modes of conduct, but not generally. Some legislation creates wealth, some reduces wealth, and some alters the distribution of wealth. The redistribution of wealth by confiscation, by the distribution of land to farmers (usually known by the misnomer of land reform), and even high death duties, all these were indirect actions, meant to reduce the gulf between the poor and the rich. They failed. Universal compulsory education cannot be enacted by the mere passing of a law otherwise the poor countries would be literate and well-off now. What sort of legislation manages to transform a nation from illiteracy by now. What sort of legislation manages to transform a nation from illiteracy to literacy? There is no general answer. There is, however, a general thesis about the success of legal innovation: if accepted and well-entrenched institutions are made to support a new institution to the point that if it fails they fail too, then it will take hold. This is an important rule. (Exceptions to it are cases of social disintegration caused by the new legislation.)

In other words, we can emulate the mechanism which Marx ascribed to the interests which he deemed primary; we can make an interest primary. For example, when we make a person's job totally dependent on his effective control of an institution, he will do all he can to control it; and if we legislate that the courts must take heed of his alarm when sounded in a special way or go out of business, then we force issues. Forcing issues is dangerous, as it may cause social and political disintegration. Hence we must make sure that only issues successfully handleable will ever be forced.

What Vladimir Illich Lenin did was to force issues beyond his control, then fail quite inevitably, and then feign success rather than admit failure and seek the cause of failure. Perhaps he feared that otherwise the system would disintegrate. The logical outcome was Stalin's systematic construction of a two-tier system: progressive but fake in the shop window, retarded but real on the inside.

Most politically active people do not know how to use political weapons. The rest tend to be overambitious, to overuse them, and to cause disaster. To avert disaster, people tend to make the system too heavy to respond to whims of irresponsible leaders, thus making political action even less possible than before. Either way, the impression evolves that there are no political weapons, only ephemeral superficial political power given to political elites to play with. Political impotence prevails and breeds utopian dreams. Political impotence and utopian dreams thrive and feed on each other. Political impotence and utopian dreams are the social parallel to personal indecision as described in Freud's theory of neurosis as the self-reinforcing futile effort to get out of a situation of conflict.

4. THE TECHNOCRAT AS A FRANKENSTEIN

We have thus arrived at the place where technocracy enters modern social and political life. Technocracy, in brief, is bureaucracy whose impotence is matched by its utopianism. The specific quality of technocracy, however, is not so much its enhanced political impotence and utopianism, but its irresponsible unleashing of physical and biological technology, which runs wild with no one around sufficiently responsible or sufficiently knowledgeable to stop it.

Mary Shelley's Victor Frankenstein has created a monster he cannot take responsibility for and so does not tame or direct or satisfy. The monster is humanized to the extent that it has needs and wants – particularly the need for love. Physical and biological technology have no need, no want, no will, no inner logic. But the inner logic of a society endowed with physical technology and no means of control of it is another matter; it does have a logic of its own, a logic that is the outcome of the logic of the situation of individual actors, especially political actors.

Who are the political actors in technocracy? Who are the people responsible for the national interest under technocracy? There may be a public administration, manned by expert bureaucrats, running the affairs of state. But bureaucrats are at best efficient as to means but not qualified to decide ends. End, then, have to be determined either by special authorities, expert or not, and this constitutes a dictatorship; or else they are determined democratically, in which case the system is an efficient democracy rather than a technocracy.

If we follow Hans Jonas and wish to compare modern machines to the uncontrolled Frankenstein monster, then we must agree with Jacques Ellul and view the machine as that of the bureaucratic, technocratic, aimless super-modern state. This picture is, indeed, favored by some foolish scientists; and science fiction fans delight in fantasizing about it – favorably, unfavorably, or otherwise.

It would be a serious and dangerous error, however, to dismiss technocracy as a political force only because it is a mere dream. For, the dream, again in accord with Marx's (and Engels's) sharp perception, is a powerful political weapon for the defense of the status quo.

Jonas and Ellul do not look like reactionaries, but according to Marx they are – Marx would have to view them as Luddites. Since the intellectual force of any modern industrialized country is enormous and able to alter our very social fabric to a very great extent and in many unforeseeable ways, it is in the interest of the powers that be, the current rulers, to keep the intellectuals politically inactive and at most ancillary to existing political activities. Jonas and Ellul are useful for them in this respect.

This point was observed repreatedly by the two great apostles of the modernization of politics in the first half of the twentieth century, Albert Einstein and Bertrand Russell. They were most strongly moved by the political needs of the world at large and by the utter inadequacy of the current political frameworks to handle the urgent problems these needs pose. They both sided with the idea of a world government – which is really but a slogan meant to emphasize the urgent need for world-wide political coordination and no more. Both these great spirits and great intellects were profoundly puzzled by the fact that the intellectuals of the world could not be mobilized to try to do something about the urgent problems of what has come to be called Spaceship Earth. The problems are now increasingly urgent, to the point that their recognition is by now sufficiently adequate for action; yet action either does not come forth or is highly inefficient and frustrated.

If anything, the urgency of the problems is by now a severe impediment to rational action. For, being ignorant and frustrated, people are ready to gamble, to follow anyone irresponsible enough to pursue a line of action single-mindedly even if it is poorly considered. It is this logic of the situation that makes whole nations follow demagogues, and it is the same logic of the situation that enables the repeated resurrection of the defunct classical liberal philosophy in Western politics. It brings wave after wave of intellectuals being towed along behind populist leadership mouthing vague pseudo-national, pseudo-democractic slogans or vague pseudo-Marxist, pseudo-socialist slogans leading nowhere in particular.

The absence of long-term policies, the mistranslation of every action meant to attack any long-term problem into a short-term activity, the frustration of all long-term intent – all this forces people who still retain some good will and who still contribute voluntary political work to harness themselves to more enlightened political action regardless of whether it is long-term or short-term. Their frustration is thus, sooner or later, bound to mount. Usually, sooner or later, they desert political action. The turnover in all voluntary organizations dealing with global issues is much too high.

We may conclude this discussion with a few simple and obvious corollaries which might be of some use. First, it is dangerous to vacillate between and optimist-quietist readiness to do nothing, in the hope that nothing need be done, and a pessimist-alarmist readiness to do nothing, out of fear that nothing can be done. Second, or the same point put differently, it is dangerous to vacillate between attending to the long-term interest which completely overlooks the short-term interests, since the short term will prevail, and attending to the short-term interest – which prevails anyway, and which may lead to disaster in the long run, and whose improved efficiency as handled by the technologically assisted administration, only accelerates the disaster. What is to be expected is the rapid growth of a public recognition of the fact that the difference between long-term utopian socialism and short-term competent technocracy is but the difference between a long-term and short-term undemocratic dream of irresponsible, competent but aimless, bureaucratic machinery equipped with modern gadgetry. What we may then strive for is a picture of a middle-range interest within democratic society which may help mold short range interest into a better frame in which some of the worst catastrophes may possibly be averted.

Furthermore, the main focus of any political plan has to be centered on the idea of political initiative. We have to determine our goal, then ask what each of us can do in order to further it, what we may expect as the outcome of our activities; and we must aim to act in such a way that under no reasonably expectable outcome will our actions be harmful – or else, if we must take reasonable risks, we must be ready to bear the losses. These ideas are simple, a matter of common sense. Their application is not. Yet the sorry fact is that initiatives are taken, often enough quite responsibly, within a number of spheres of life though seldom in politics. This is so for a number of reasons that may very well be taken into account. First and foremost, one cannot take initiative without having a plan of action, and in political life, for whatever reason, plans for action are scarce and most programs of candidates for office and of governments of all sorts are hardly more than general sketches. Second, whereas most rational action is executed within a social framework and is constrained by given sets of factors, rules, and circumstances, politics often has to handle the framework itself. This is true of wars as well, except that the logic of a war is infinitely simpler than the logic of politics. (Wars are more often bungled than any other activity. This is getting exceedingly dangerous these days when generals are in charge of so much and such potent destructive machinery.) Also, of course, the two difficulties mentioned strongly interact with each other: it is all the more difficult to have a global political plan. Furthermore, the attempt to change the political framework means that what is offered is not explained to the public because they might resist it; a possible result is that change imposed by a revolution might fail either because it was poorly conceived or because the public objects. And a failed revolution may strengthen conservative feelings, thus impeding change, and the forces of change may then need to seek a new revolution]

What characterizes a stable democracy is two things, the flexibility of its framework, which permits changes within it, and, second, a rational way of changing the framework itself. This is the idea of constitutional amendment. The very stability of democracy, the very ability it has to absorb changes, calls for some institutional means of regularly reassessing the system as a whole.

All this should make it clear that the very discussion, by Jacques Ellul, Hans Jonas, or Herbert Marcuse, among others, of the aimlessness of technological society is itself aimless. It cannot lead anywhere, and the situation cannot be improved unless we discuss technology within a concrete society, unless we decide what are the given goals of that society, its decision making apparatus, and so on – about all of which these writers are extremely cavalier.

A society, even mankind as a whole, has a few given goals: survival and what must go with it, and perhaps anything that can make survival worthwhile. Different political life styles compete for the position of that aim which might make survival valuable: democracy, perhaps anti-democratic socialism, perhaps democratic socialism. Once these basics are spelled out, the fears of an aimless Frankensteinian technological monster must be replaced by the quite different fears of an aimless political monster. But given the ideal of liberal democracy within nation states, an assessment of the current situation within Western democracy becomes urgent. An image of democracy for the not-so-immediate but also not-so-far future is required. We have gone too long without a proper reevaluation of our international and national politics.[11]

5. CONCLUDING REMARKS

By heightening awareness of the state of current political philosophy and political science, by attempting to make use of the best ideas and intellectual tools available, we may achieve a new situation. Assessing the traditional frameworks shows their defects. Assessing such progress as has been attained despite these shortcomings is less easy; doing it may be the opening of new modes of thinking, may lead to the development of new intellectual frameworks for political thinking that will take better care of technology.

Perhaps the most important pervasive fact to observe is the fuzziness of the situation. Political parties are meant to be instruments for developing political plans and political action. Yet as soon as parties develop they are prone to function not only as political institutions but also as social institutions, offering safety and demanding loyalty. And no sooner is an economic concern developed, like the automobile industry, than it becomes a major political factor and governments are forced to save it from financial ruin rather than face the massive unemployment that might ensue. The deep interactions and connections between politics, society, economy and technology cause today new types of problems calling for a new type of integrative solution.

The division of labor, together with the identification of the economy with technology and of technology with science constitute a major incentive to specialization. The position of an apprentice, consequently, has changed very drastically, from that of a young worker open to diverse influences who is challenged to learn diverse skills, to that of a young super-specialized assistant to a specialized master. Clearly the status of an apprentice is more legal than social. Governments might invest less money in interdisciplinary research institutes, which are today so fashionable, because they invite galaxies of specialists from diverse fields to bask in their own glory. Doing this would allow these governments enough money to facilitate the growth of new types of apprenticeship – including retraining and further education and the evolution of new kinds of interdisciplinary training that are not second-best to specialization. And this could be protected by simple and modest and inexpensive legislation that could easily be enacted by tying it to current grant monies, tax exemptions, and the like.

The situation is doubtless very gloomy and it looks as if we are going to lose the whole fight. But in the meantime there are so many simple things that can be initiated which might open up new longer-range options that the absence of political initiative is barely understandable. The main line of action should be, obviously, to initiate moves which will raise the likelihood of public debates about the central and difficult problems of the day, in which further proposals for further initiatives might possibly be offered by diverse participants.

PART THREE

TOWARDS A CRITICAL PRAXEOLOGY

INTRODUCTION TO PART THREE

The previous two parts of this volume constitute attempts to present an integrated image and an integrated critique. This part, all too brief, aims at an integrated proposal – not so much of a detailed theory but of a sketch of a program for action. The first half of this part is devoted to a general idea of programs for action: the need both for medium-range goals and for means to integrate them with current, operative, short-range ones. The second and concluding half of this part is devoted to questions regarding possible implementations of desired major changes as rapidly as possible. The shortcomings of this last part of this volume are painfully obvious. It constitutes an invitation to invest new efforts with hopefully more interesting and encouraging results.

CHAPTER 11

AN IMAGE OF A BETTER FUTURE

There is a traditional disagreement between utopians and pragmatists in political thinking. The utopians from Plato to B. F. Skinner say that when we describe our ideal on a large canvass we have a better idea of what we are after. The anti-utopians from Spinoza to Popper want something more immediate. The conflict between the two stems from the fact that the anti-utopians declare the utopian image harmful. They say it makes us ignore the imperfections of existing society, especially the need to make room in our institutional setup for human frailty. This has led the anti-utopians to propose the idea of piecemeal social and political engineering, as Karl Popper has put it, on the ground that the pressing problems of the present must be on top of the agenda of the planners of a better society. Nowadays, when the urgent problems are global (poverty, population, pollution and nuclear proliferation), the distance between schools is shrinking. We may imagine, as a guideline or a guiding thought, a world in which these pressing problems are solved, yet which is not in the least perfect and which therefore still makes allowances for human frailty and the need for further progress. Let us call such an image a near-utopia. The present half of this concluding part of this study is devoted to the image of a near-utopia of the kind outlined here: democratic, problem-ridden, but with an effective control of population and pollution and nuclear weaponry.

1. TOWARDS A THEORY OF PRACTICAL PROBLEMS

The logic of questions is a young field of inquiry, and it seems to have as yet almost no field of application beyond certain very narrow fields of computer technology. Nevertheless we have to glance at the way problems are made agenda and at the way problems are given priority on the agenda. We often overlook even such a simple and obvious fact that problems with sufficiently low priority are never discussed even when put on the agenda. People inexperienced in the democratic process are repeatedly defeated by their oversight of this fact; they have invested much effort in their attempts to put certain issues on the agenda, seem to have won, only to be frustrated by the discovery of a defeat by default with no appeal. Witnessing such a

213

frustration and the resultant despair of democracy is exceedingly unpleasant. No less unpleasant is the discovery that an irresponsible leadership allows a debate to spread to the utmost in order to defeat a motion by default due to lack of time for a proper vote on essential issues. This is why a training in the democratic process for a wide public is essential. The study of the logic of problems may also improve such training.

The logic of scientific questions has preceded the logic of questions by over one hundred years. It was Dr. William Whewell who said, early in the last century, that since unification is the supreme task of science, researchers are constantly faced by the same two root problems: how can we unify some of the facts and theories of science into new theories? and how can we test the new theories?

Even in science things are not that simple. In its effort to comprehend the universe, science goes both in the inductive and the deductive direction – explaining the facts as Whewell described and also offering intellectual frameworks for these explanations to fit into. Yet the moves from the two directions which should capture the task and encapsulate it do not quite fit. And then the problem arises, how can we fit the two? In addition to this, problems of priority abound – which problem comes first, which part of a theory in need of modification should one try to modify first, should one improve one's ideas or one's apparatus, mathematical or experimental, etc., etc.

Nevertheless the sphere of scientific problems is miniscule and highly ordered in comparison with technology and legislation and social organization; so much so, that one can hardly know where to begin in the discussion of problems in general. We may, then, proceed from what is easiest to present – not necessarily easiest to practice.

The simplest kind of problem is that of selecting an item which satisfies certain criteria from a list of items each characterized by a given list of criteria. Think of questionnaires where each person has a name, date of birth, address, occupation, etc. Each characteristic sifts from a list of names the names of individuals sharing it, and ideally a question can be, who has such and such characteristics? Also, how many members of a given list share a given (set of) characteristic(s)? And so on.

This kind of question is most specific and is best handled with the aid of computers. Purchasing an airline ticket with the aid of computers, taking stock, deciding on a purchase, storing, etc., are all highly sophisticated affairs of the same kind as far as the logic of questions is concerned. Mathematically things may be different, and new branches of mathematics may be developed (such as queuing theory and (linear) programming) to handle matters; but once handled, they enable us to handle questions with the aid of computers.

Next come questions within a fixed intellectual framework, yet not given to computer treatment since not formalized. Most everyday problems we meet are of this kind. We have an intuitive feel which is hardly controverted by our peers as to what constitutes an answer to a given question, and what would constitute an outlandish answer.

The most important thing to notice about such questions is that at times they are controverted, at times there is a consensus about the right answer to them, and at times they are unsolved, having no known reasonable solution. But this most important point is regularly ignored by a standard and very widespread confusion of the word "solution" with the much more stringent "true solution". Thus, usually, the sophisticated undergraduates are just as unsophisticated as the man in the street when they have to answer the question, "considering the question, 'what day of the week is today?' how many answers are there to this question?" They think the true answer is "one", when obviously the true answer is "seven": they confuse "an answer" with "the true answer". Obviously, the confusion is here easy to make since the question is not controversial. Things are worse when the question is controversial, and in the following way.

When we have a controversial question, we usually have some uncontroversial answers to it. This sounds odd, since when an answer is uncontroversially endorsed, the question it answers is uncontroversial. But this is not the case when an answer is uncontroversially rejected; and it may be unanimously rejected on various grounds, especially as obviously false or as fantastic. Consider a controversial question. The confusion mentioned before now narrows down. Suppose we controvert the question, "what day of the week is today?" yet agree it is a work day. How many answers would we say it has? It still has seven answers, but confusion still blocks this true answer. People may still confuse "answer" with "true answer" and say, "one", or, they may add the following to this confusion. Not knowing the true answer (since this time, we assume, the question is not uncontroversial), people may confuse "answer" with "possible answer" or "candidate for the status of a true answer"; in this case they will offer a number bigger than one and smaller than seven.

In other words, when deliberating on an open question we tend to exclude a priori the obviously false answers to it. This is a serious error in all but everyday cases. What characterizes an everyday question is its ordinary setting, and it is ordinary settings that make some possible answers obviously false. This is observable when a new technology becomes widespread: it makes certain obviously false answers possibly true. (Think of a question about the source of a given item of information and its relation to the growth of telecommunication technology.) Hence, the best way to setting the (everyday) background to a problem (which is what makes the problem everyday) is to present all the obviously false answers to it and give the obvious refutations to them. The advantages of this technique are so obvious that they will not be stated here. Instead, it might be noticed that at times the false answers may be presented in an obvious order.

Things get exciting when we translate a question from one framework or system to another (this breaks away from the everyday). We may ask, what minor and reasonable variations in the framework or system may revolutionize the list of possible answers to our question? This kind of question is an invitation and a challenge to inventors. It is quite possible at times to present this challenging invitation in a new system and to apply to it what has been said thus far. When it is possible, we have hierarchies of frameworks or systems. We also have then hierarchies of questions. We may now observe the following very useful point about the two hierarchies.

Very often, when a system broadens, it pays to search for a way to broaden questions accordingly. Many technological problems were never solved because they were broadened and the broader variants of these problems found other, ready solutions. Every time a technique is revolutionized, whether by replacing the horse with a motor or replacing the piston engine with a jet engine, or even when replacing standard arithmetic with nonstandard arithmetic (Abraham Robinson), a vast number of problems cease to be pressing, though their successors are broader problems which, in their turn, generate new specific problems.

Hence, sticking to specific problems, especially in technology, may be sheer waste of time. And the history of technology is full of ingenious inventions which were outmoded before they reached the market. (The same applies to premature inventions.)

Moreover, the more one gets stuck within a framework, the smaller one's problems tend to get. This is true even of one's habits of everyday conduct. One may like this fact and one may rebel against it. Socially, however, the more cumbersome it is to move away from a framework, the less one should expect to have a solution which breaks away from it to enter the market. Examples abound. The reform of the typewriter key or of the way to write music (especially for the cello) are paradigms. Only when a solution is spectacular (such as having one key-apparatus geared to many keys by preprogramming) it may cause a major change. As Imre Lakatos, the first student of problem shifts, has noticed, the construction of expensive scientific instruments tends to narrow the range of scientific questions acceptable by the profession and so to prescribe a degenerative problem shift. This is a warning worthy of attention, especially when making large-scale investments in instruments and in calculating future benefits: one should consider the possibility that the instrument may very well be quickly obsolescent. (Think of the micro-computer).

The last discussion concerns frameworks: which frameworks are possible within which to solve certain scientific and/or technological problems? Here we move to all sorts of fields of activity, from metaphysics to science fiction.

Once we look at an arbitrary hierarchy of kinds of questions, as the one presented here, beginning with fixed sets of answers to choose from and ending with the search for new metaphysical presuppositions, we can very quickly see how very different kinds of questions may interact: how much should the framework alter before it becomes advisable to reprogram our computer which stores sets of possible answers? Which scientific theories deserve rewriting in the light of new presuppositions? Is it worth trying to write quantum theory in the language of general relativity? Is it worth trying to quantize general relativity? Or parts of it? (Yes: the relativist black whole theory was quantized.) Is it worth using general relativity in aerodynamics? (No.) In navigation? (Yes.) Should scientists worry about every empirical achievement not scientifically explicable?

The logic of questions is in its infancy; the idea that a problem shift may be progressive or degenerative is new: Lakatos took it for granted that it should always be progressive, so remote was he from technology. The fact that questions are discovered was discovered by Michael Faraday over a century ago, yet this remains unknown.[12] The fact that questions are often left unsolved due to a shift in interest and/or circumstances is fascinating too; it has been observed by some historians of science, but has been left unstudied.

This is a fertile field of study: the logic of questions and its different applications to scientific research, to technological research, and to practical affairs (technology proper). In the field of education it may cause the long needed revolution: teach students how to handle questions and they will be free.

It is necessary, in conclusion of the present discussion, to mention cost-benefit analysis. Cost-benefit analysis is the evaluation of a venture into a given possible investment and the ill side effects of the venture as cost, and the profit and beneficial side effects of it as benefit. These things cannot be evaluated: we do not know the cost and profit of anything ten years hence. The way to compute cost, or benefit, into the future is to compute at a discount-rate: how much should we pay now for an expected commodity in a given future whose cost now is given? The cost-benefit analysis of an underground train line in London, we are told, was deemed negative, and advocates of it changed the verdict from negative to positive by convincing people to raise their estimate of the discount rate of the distant future of a few decades. What can be said of such a procedure except, perhaps, that it is cheating? It is, incidentally, hardly cheating when one does one's cost-benefit analysis in different ways and observes which way the outcome is positive, which negative. What is the virtue of doing this exercise?

The problem looks difficult because we started with cost-benefit. The starting point is not cost-benefit but opportunity cost. The problem of opportunity cost is much simpler: what options do I give up when making a given choice? As everybody knows, getting married has the opportunity cost of leaving the marriage market altogether – for life or for a while, depending on many factors. What does one give up when one purchases an automobile? Usually not food and not even clothing, though perhaps some reduction of clothing quality may be called for; and, of course, one saves the cost of alternative transportation but has to pay for running cost and maintenance; and so it goes. It is only because cost-benefit analysis is a means of assessing opportunity cost that a crude use of that method may be at all enlightening.

A caution: opportunity cost is only meaningful on the hypothesis that one is utilizing all one's given resources. The moment the choice of an alternative also increases one's resources, things look different, of course (as in the case of marrying a rich person in order to have money to find favor with another: a standard soap opera theme). This is why national cost-benefit analysis makes better sense than cost-benefit analysis for a firm. Even so, we hardly know what are all the options, and only when we eliminate almost all options on the basis of a crude estimate and wish to have a better view of a few of the remaining options, is it reasonable to do more precise estimates and a comparison, and when these estimates are numerical they are cost-benefit analyses: they should be as comparable in rough assessments of discount rates as possible.

The intermediate stage, prior to the drawing of the cost-benefit analyses, can be effected to varying degrees of systematicity and precision, and along diverse lines of thought. On this technical matter there is a vast literature, which is usually much harder to fathom than need be – for want of a framework which may assist one's efforts to see the worth of the techniques one is invited to acquire.

The present discussion is a contribution towards the construction of such a framework.

The guidelines for the logic of questions in general, and for the logic of practical questions in particular, then, must be two:

(1) There are reasonable and unreasonable, or intelligent and unintelligent disagreements; the reasonable or intelligent disagreement is invited when there are two or more reasonable or intelligent answers to a question.

(2) The difference between the reasonable and unreasonable depends on general presuppositions.

The chief corollary from these is this:

It may be very useful to render one's general presuppositions explicit and examine them on occasion.

2. DEMOCRACY, DIALOGUE AND RESPONSIBILITY

There are many things which cannot be done in any reasonable democracy, which can be done in other sorts of political systems, and which are at times highly desirable. (Democracy, too, has its opportunity cost.) The paradigm case for this is the ancient Roman law enabling the senate to appoint a dictator for six months in times of emergency. In principle this is not much different from the kind of coalition government which Great Britain had during World War II. All that remains for the critics of democracy to prove is that emergency is the rule rather than the exception, and farewell to democracy.

The force of this argument is tremendous. It is the recognition of this force which made Pericles declare, on the site of the open graves of Athene's war heroes, we are not suspicious of our neighbors.

We are not suspicious of our neighbors. As an assessment of the situation this pronouncement of Pericles is insane under the circumstances; but as a matter of policy it is both noble and wise: we choose to live well rather than in a state of preparedness. How true: the Nazi slogan, cannons instead of butter, can very well be reversed – instead of cannons, let us produce milk for our children and bread for our workers. It sounds irresponsible, and it certainly can be; at times it was just that. Yet the opposite is also at times quite irresponsible, and more often the irresponsible political leadership preferred cannons instead of butter to milk for children instead of cannons. Moreover, the need children have for milk is often more obvious and more immediate then the need a nation may have for cannons. Furthermore, a high standard of living is much stronger a military weapon than cannons, as everyone can see – even though defense is still necessary, no matter how good the society is.

What we have before us when discussing the pros and cons of a political regime is a systematic error which we do our best to perpetuate throughout our educational system – so that the survival of our democratic system, as Samuel Butler has observed (in the conclusion of his exciting *The Way of All Flesh*), is due to the imperfection of our educational system – of teaching people to defend the views they hold and attack the views they reject, to defend the actions they choose to take and attack the alternative actions which they reject. It is clear to anyone who puts his mind to it that a responsible attitude calls for checking, and that checking is considering the possibility that one is in error, and that considering this means examining the best arguments available against one's own views and for the opponent's views. It is a disturbing empirical fact that when one says this thing one meets with impatience. The impatience is, to continue the empirical observation, defended by the claim that the point conveyed about the need to examine opposite views is either a trivial and well-known statement or a thinly veiled homily or both. And perhaps this defense is true. Nevertheless the empirical fact remains that even in highly civilized society the habit of defending opposite views is extremely rare. What is to be done about this?

This is a question concerning implementing standards or implementing reforms of standards or at least improvements of current standards. Certainly standards are high in some quarters. Civilized courts and civilized committees – on diverse levels of government, of public administration, and even of voluntary associations – are run on the principle that opposing views should be both defended and attacked. Yet these facts, to conclude the empirical observation, are not appreciated even by scholars, let alone by the general public.

The striking example is the proverbial question, have you stopped beating your wife? The question is generally viewed as unfair – because it implies that you have for some time been in the habit of beating your wife. Unfortunately writers of the logic of questions endorse this view, which is very crude, on top of being erroneous. Almost any answer one gives to almost any question may be read in a manner that puts the person who gives that answer in some unfavorable light, as writers of courtroom melodrama love to illustrate with a never-ceasing ingenuity. And even the most unsavory answer may hide a noble intent, which is a favorable point of tellers of twisted tales. What the question, have you stopped beating your wife, is supposed to illustrate is that ordinarily ordinary citizens need protection against the ascription to them of some malice. This is not true; it is a fundamental and sacred principle of democratic society and of the democratic way of life that every individual has the right to ascribe any motive to any other individual. Normally, also, it is the duty of the attacked party to remain unprovoked and to avoid any act of self-defense. At times an individual needs, however, to have the opportunity to exercise the right to dispel ascription of base intents – particularly when he is the accused in court. It is then the duty of the court to see to it. If he knows that, and if he trusts the court to discharge its duty, then he will have no fear of the malicious interpretation of his answer. The council for the prosecution asks, have you stopped beating your wife? and the accused says, no. Then it is the turn of the council for the defense, whose job is to ask, why have you not stopped beating your wife? Because I could not, is the proper answer. Why? Because I have not started beating her. Jury and audience smile in relief. End of episode.

The empirical fact which seems to endanger democracy most is that even under democracy individual citizens feel repeatedly (1) attacked, (2) in need of defending themselves, and yet (3) not having their right for self-defense properly safeguarded. Especially as long as they are not properly accused they justly feel that their right to defend themselves is violated. It is then their duty not to attempt to protect themselves, since they recognize the court which sits in judgment over them, yet one has the duty to refuse recognition to any court not constituted in proper consideration for their rights. (This duty is easy to discharge in a democracy, but not otherwise – particularly not where inquisition methods are practiced, whether in Renaissance Spain or in modern Soviet Russia.) By allowing others to judge us we tempt others to violate the commandment, judge ye not! and become accomplices in the violation. Moreover, allowing others to judge us without first safeguarding our right for proper defense, we allow others to act against us arbitrarily and thus we tempt them to blackmail us: do this or that for me or else I judge you lazy or stingy or heartless. It is an empirical fact that in democracies this is the commonest mechanism of forcing people to conform and to act against their (and the public's) interest.

As a hypothesis we may correlate two widespread modes of conduct, the erroneous acceptance of others as properly constituted courts of sorts and the erroneous readiness to defend one's own views and attack others'. Within proper democratic procedures the two errors cancel each other. In court each party is supposed to advocate one view, and on the hypothesis that the court will hear both the defense of and the attack on all the views advocated by the different parties. Unfortunately, often one is oneself the court yet one behaves like a defendant. For, one deems oneself to have acted as a court, made up one's mind, and is now accused of having decided erroneously. One forgets that even when a court-of-appeal overturns a judge's verdict, it seldom accuses the judge. Otherwise, if and when he is thus accused, then he has the right to defend himself – not during the appeal session, but quite separately.

This explains the discrepancy between people's perception, self-perception, and conduct. When they say they need not be told we should weigh the pros and the cons of competing views, they mean, in truth, they know this rule and, indeed, follow it whenever possible. Yet, usually enough, they think each of us should do so in the privacy of one's personal thinking, and then come out with a reasoned opinion. This reasoned opinion may publicly clash with another person's different reasoned opinion in the manner usually followed in court: each side defends their own position and attacks their opponent's opinion, so that judge and jury – the general public – are presented with pros and cons of each alternative. Yet when one's opinion is a minority opinion, spokesmen of the majority opinion might challenge the holder of a minority opinion in a manner which looks like an accusation: the holder of the minority opinion is accused of not having done their homework: were they more familiar with the relevant information – as they should be, before they dare publicly voice a dissenting opinion – then they would have seen that their opinion is erroneous. Moreover, either the holder of the minority opinion has some privileged access to some information, in which case it is their duty to divulge it and make it public, or else the holder of the minority opinion is challenging the holders of the majority opinion, and also in a manner which looks like an accusation: the holders of the majority opinion are accused of not having done their homework: were they more familiar with the available relevant information – as they should be, as responsible spokesmen of the majority opinion – then they would have seen that their opinion is nothing but a popular prejudice. Moreover, etc., as in the previous sentence or two. The cycle of accusations and counter-accusations tightens. The classical example is the debate about phlogiston where Mme. Lavoisier ceremoniously burned some of the best books ever written for no better reason than that the doctrine they propound is false.

The accusations are launched within a fake framework: one feels accused as in a court yet there is no court. Why does one feel that way? This may be a psychological question to do with a misperception. Let us ignore the theory of perceptions and misperceptions. It may, alternatively, be a psychological question concerning one's sense of moral autonomy and one's sensitivity to any assault on it. This is true.[13] It also indicates that much more education for autonomy and for hygiene concerning one's autonomy is required in our society. This calls for a deep reform of our education system, in which training for autonomy should be a major item on the educational agenda. Looking at the question sociologically, however, one may often see in the procedure described her – kangaroo – court, as it is called in the vernacular – a court informally constituted by an informal professional guild. The professional guild at times has the right – the medieval kind of right – to judge its members as not living up to the guild's standards. When one is properly accused of ignorance by a guild court, the ignorance may be both rooted in irresponsibility and the root of one's not accepting the views widely accepted in the guild. In this case it is hard to dismiss the accusation, even though formally recognized guilds and guild courts are medieval institutions which should be abolished as undemocratic. Nevertheless, informal guilds and informal guild courts are in this respect worse than formal ones.

It is thus a mistake to take up a challenge to defend one's views. It is a simple legal and moral fact of democracy that everyone has the right to hold any opinion whatsoever. Yet when one holds a minority opinion one is often challenged to a verbal duel – the challenge being not intellectual but moral, and therefore one which should be rejected with contempt until it is reissued as an intellectual invitation rather than as a moral challenge – not a duel with weapons but a chess game.

This sounds both very satisfactory and very unsatisfactory – to report yet another very important empirical observation concerning all sorts of intellectuals in the second half of the twentieth century in diverse parts of the Western world, students and professors, professionals and amateurs. The satisfactory aspect of it is easy to spot: it is the constitutional democratic freedom of opinion, of conviction, of persuasion. This freedom is admittedly not always clear and uncontested. Yet, in democracy, we all wish to defend, as much as reasonably possible, anti-science and irrationalism and even intellectual irresponsibility. It is, we agree, politically permissible though morally forbidden to be irrational and intellectually irresponsible. Hence, admittedly, there is no room to challenge our political right to our opinions. Yet we may still invite a moral challenge to argue that our holding our opinions is not as morally irresponsible as that of the anti-rational. Should we not prove ourselves morally responsible in our being reasonable and rational? This question – or rather the accceptance of an affirmative answer to it – is the root of the dissatisfaction people have with the content of the previous paragraph.

They are in error, and their holding of this error is their way of accepting the challenge of holders of opposing opinions as moral challenges and hence as accusations.

To simplify matters, let us assume that the differing views are expressed as differing modes of conduct, so that an eccentric mode of conduct is challenged as irresponsible. It is harder to view eccentric opinions rather than to view eccentric conduct as irresponsible. Indeed, when asked for an explanation, the accusers of the holders of eccentric views refer to the conduct of the holders of eccentric views, such as their confusing or mis-leading the public, such as their corrupting the youth of the country, and such as their spreading error and lowering scholarly standards.

When is eccentric conduct reprehensible – as irresponsible or otherwise? Only when conformity to the majority mode of conduct is obligatory. When is this the case? In an emergency, for one thing. Or in cases where, if anyone does his homework reasonably well, one cannot avoid the conclusion that the mode of conduct adopted by the majority is the only right one. Hence, when two disagreeing parties accuse each other, they imply that there is only one responsible solution to the problem about which they disagree.

Most problems which are debated in reasonable company are open to more than one reasonable solution, and, surprisingly, also often to more than one permissible solution. The paradigm is the so-called paradoxes of the calculus of probability which at the end of the nineteenth century baffled some of the best minds – mathematicians, logicians, and philosophers. When a problem has more than one solution, but looks as if it has only one solution, then one is baffled by the very presentation of two solutions plus the proof that both are permissible. In practical life this occurs much more often than in mathematics, since there we often meet solutions to problems which seem to require unique solutions.

The classical theory of rationality took it upon faith that every clearly defined problem has one acceptable solution and every clearly defined task has one rational execution; alternatively, that the class of acceptable solutions is well demarcatable in a manner clear for all to see.

Scientific technology is particularly open to alternative permissible solutions, since in technology a solution is judged permissible when it passes standards of responsible implementation as specified by law. Thus, no one is forced to offer the service of air travel or accept it, yet also no one is allowed to offer that service unless licensed. The licensing is a complex operation, which includes the certificate of airworthiness to any given type of vessel. Before the certificate is given, certain test flights are obligatory; test flying is certainly not obligatory: it is a highly paid risk operation. Nor is irresponsible test flight permitted: no vessel is permitted to be test-flown prior to certain ground tests, such as wind-tunnel tests.

The rules of responsible conduct are not a matter of personal conviction. A pilot who feels confident and is willing to test a vessel prematurely is not permitted to do so, and one who feels diffident is replaced by a colleague.

Yet this is not to say that the rules of responsibility are perfect. We may be deemed irresponsible if we thoughtlessly hide behind rules despite their obvious deficiencies. In the age of danger to the ecosystem we all know that all systems of rules of technological conduct are highly unsatisfactory. We do not know, however, what are the rules with which to replace current rules.

We may seek these responsibly in current debate. Debate, however, will flourish only if we realize that we may responsibly dissent from current public opinion without needing to defend our opinions when challenged. Unless we see that, we are stuck. Once we see that, myriads of new problems arise. These problems may or may not be solved – they may even be insoluble.

Yet we must begin with the assumption of a bona fide: we are not suspicious of our neighbors. We must have standards of responsibility, but they must be rather low. And we must assume that the rational is the permissible, not the obligatory, and hope that we wish to act rationally. And we must assume that dissent is rational and responsible except for the cases in which we are forced to conclude to the contrary.

This is a necessary basis for any sincere effort to encounter our current formidable global problems.

3. AN IMAGE OF A HALF-WAY UTOPIA

Utopianism is objectionable as it eliminates disagreement. It is also vague. The utopian dream is a long-ternm project, whose immediate impact on day-to-day political and social and economic decisions cannot be clear. This is true even of the question, how will we have solved all problems of scarce energy in our utopia? It is not at all clear that saving energy today will help us come nearer to that distant goal, nor that wasting energy will. Arguments go hither and thither and a priori more arguments support the view that the way we utilize today's energy resources will have no effect on the distant future. The energy resources of utopia will be utterly different from ours and quite unproblematic. It is particularly hard to know or imagine the values and tastes prevalent in Utopia. To imagine that our descendants in Utopia will have the same values and tastes as we do is naive, simply because we do not know if our value system is consistent even in the present, let alone under unforeseeable external conditions.

Consider the recent sexual revolution and the changes in values and tastes that it has introduced; consider only that part of it required by three very important factors: first, the equality of women; second, the need to control the size of world population; and third, the need to educate one's offspring in a manner not available without some family stability. Now this little is hardly controversial; it is also too complex. It will change unpredictably our ways in different spheres, including the sphere of work, and in the sphere of the family. Women who can control their sex life and reproduction can seek equal employment. And as the family is deeply linked to employment, the change in women's occupational work roles will have to go with changes in family structure and domestic and child-care work roles.

Less powerful considerations than these have led to the anti-utopianism of the early post-war period, led by Karl Popper, and Daniel Bell, among others. They concluded that short-term political problems should suffice to engage us. The term Popper used is piecemeal social engineering, where piecemeal does not necessarily mean small, though it usually should, but rather short-term, beginning always with the most urgent problems on the agenda.

The complaint that piecemeal social engineering alone may cause a loss of general orientation has been aired. In response, Popper insisted that the tendency to relieve misery and increase the personal freedom of the citizen should suffice. It does not. The increase of all citizens' personal freedom, or the increase of most while compensating the rest, has proven impossible even under idealized conditions (Kenneth Arrow). The increase of freedom of some at the cost of reducing freedom for others is quite unacceptable.

I. C. Jarvie has proposed that thinking up utopias should be encouraged as a means of inspiration and as an aid to the shaping of goals. He is right and even a stronger argument is valid, which makes his proposal not only advisable but even necessary: we can make do, with or without some utopian thinking, only when we have medium-range goals that are reasonably easy to achieve consensus for, and today we have no such consensus.

Consensus is the key: social and political action require consensus – national or international, depending on the scale of the action. One may expect that a minority in the nation will oppose any given plan, but a national consensus may be achieved nonetheless. Indeed a consensus may be achieved even when a majority of the nation does not agree yet approve: even when the vast majority of the nation have their doubts, a consensus may be achieved because the people are willing to let a leader try out an idea he has – and at times this is rational (Churchill), at times not (Khomeini).

We need a different kind of consensus for the purpose of determining medium-range national-political goals. Once we determine these, we can ask what part does/can technology play in a society of the kind we wish to aim at.

There is no such consensus anywhere on earth. Yet we can all come together and ask concerning what kind of goals is it reasonable to try and attain a national consensus. This is a different kind of inquiry, subject to a different kind of inner logic. Some of the people engaged in a discussion of this kind might envisage, for example, some very liberal code of sexual conduct, and others will envisage, perhaps, a very strict code. It will not be hard to agree that such a matter will not easily reach a consensus now, nor that it is urgent. Perhaps one might argue that the semi-utopia we may wish to develop a consensus about will allow a pluralism of sexual codes, but even this is too problematic and we may wish to develop our image of our semi-utopia without too much deliberation about sexual codes.

Other matters will be much harder to ignore. Since current jails are schools for crime, it is clear that they have no place in our semi-utopia. Utopia proper has no crime: alternatively it may have almost no crime plus some simple techniques for handling criminals, such as lucrative exile. Semi-utopia will either need much better jails, with advanced physical technology for their control, or no jails at all – as seems reasonable to all enlightened criminologists. It is hard to see how quickly a consensus on this matter is attainable, and it is hard to see how this matter can be avoided.

And so an image of a semi-utopia should emerge through a public debate among concerned citizens, not on the supposition that it gains consensus thereby, but on the supposition that it is both worthwhile and reasonable to assume that it is possible to achieve such a consensus.

There will be no detailed proposal here for such an image; only a contribution towards it. For example, it is obvious to all that the semi-utopia will not be and should not be outright anti-technological. It almost follows that the use of computers will then be no less common than the current use of the typewriter among college graduates and journalists. This already makes the communication system of the semi-utopia much more powerful than today's, as today's is in comparison to what it was before the invention of the telegraph.

We must assume then that all exams will be conducted with the permission to freely use one's computer terminal. The changes which this will necessarily implement in the educational system will of necessity be enormous. Not only will memorizing be killed: students will be able to plug into the best lecture in the country so that the senseless custom of frontal lectures will be overcome at last and be taken over by public performances, by small class activities, by exercises, and by workshops – perhaps only workshops, supplemented by discussions and work groups of two or three participants each.

The liberalization of student-teacher relations, or even of pupil-teacher relation, will at once improve parent-child interactions.

Quite generally, what we may see in our semi-utopia is a broad acceptance of the liberal and the democratic style and principles, so that prohibitions will have to be justified and so that democracy will live on diverse levels and in all organizations. Consequently, we shall find an enormous diffusion of roles, of public participations, of functions. It will become increasingly difficult, it may be hoped, to distinguish teacher from student, doctor from patient, worker from visitor or from manager; it should become increasingly difficult to distinguish a class from a workshop from a union meeting from a working party. It should be increasingly difficult to distinguish a work place from a center of learning and both from a place of recreation.

The need to distinguish friend from foe, brother from stranger, resident from visitor, these needs are very strong and deeply ingrained. They are rooted in fear and insecurity. The growth of security and education may reduce them. When a culture is strong – be it a sub-culture, a religious sect, or a national culture – there is no worry about those who leave and no worry about those who join and no worry about those who stay. When a culture is weak, when all of a sudden a culture is concerned with its own survival, with the survival of its peculiar traits, with its continuity, then those who leave become deserters, those who join become a threat of dilution or of takeover. Once we realize this, we may ask, what is that which merits preservation and development and why? Should a weak culture be preserved?

Such an approach does not in the least guarantee the preservation of that which deserves preservation. We may be mistaken about what merits preservation. We may be mistaken about how to preserve that which merits preservation. We may be lost together with whatever we are trying to preserve, either because we are unwittingly handling explosives or because our neighbors do, or because tomorrow the sun will become a supernova. But in the mean time rationality is, as best we know, still the order of the day.

It is not possible to conceive of a semi-utopia without some view of foreign relations. In the current situation the same holds for global policy on the ecosystem and on world peace. We must, however, begin with the internal structure of our semi-utopia in order to see how much technology it contains and what kind of technology it has.

The growth of physical technology of necessity will bring about more social change – it will have to create a new social technology. It is clear that in our semi-utopia it will be easy to decide to change one's job or one's working hours or the amount of time and effort invested in one's job annually. This has to invite a new kind of labor organization, and a new kind of public administration – especially of emergency services. And at least one thing must be unquestionably accepted in the semi-utopia: we must eliminate current blatant injustices concerning the assignments of jobs and the remuneration from work – especially race and sex discrimination and segregation. Also, today's various versions of the work ethic will be extinct. Most societies recognize the right to work only minimally and our society recognizes also the right to full employment; the semi-utopia will uphold both. The incentives to work should suffice, and no compulsion of any kind will back it, economic, social or moral.

The condition is almost achieved in advanced technological societies, where both early retirement and public welfare are common. Welfare officers harass welfare recipients – on the erroneous assumption that the quality of working life cannot be improved so as to make work more attractive than welfare for sufficiently many people to prevent welfare from being an impossible burden to the economy. Hence the harassment is a disincentive to the efforts to improve the quality of working life.

But let us leave these details, stay with the general idea of a society with no more social pressure and constraints on individual citizens than deemed necessary for the survival of that society, and move to the much tougher area of international and global policies.

4. AN IMAGE OF A PEACEFUL WORLD

It is very easy to describe what is the barest minimum required for the stability of the globe. It is much harder to know if this stability is at all attainable, let alone the required international agreements on a course of action before any action can be meaningfully affected. What the world needs as soon as possible, is control of its population, of its environment, and of its stockpile of nuclear weapons; and it needs the closing of the gulf between rich nations and poor ones; and it needs political toleration. A tall order, but it cannot be shortened.

Our semi-utopia must maintain relative peace in a world relatively devoid of abject poverty and of political intolerance in a relatively efficiently and globally controlled environment and population. The road to such a situation is hard to envisage; even some of the most basic features it involves are too problematic as yet. The military aspect is the easiest to envisage, and even the major features of the way to it are not hard to imagine: de-escalation of tensions and increased trade and other normal exchanges should gradually lead to almost total disarmament. The trouble lies in the normalization of international relations which is essential to the stabilization of each stage of the process of de-escalation leading to the next. Normalization has thus far taken place only within the industrialized democratic world and for the simple reason that inner stability permits an openness precluded by the present inner instability of the less fortunate countries.

Hence it is in the interest of the industrialized stable democratic countries to have the rest of the world well-off and stable (John F. Kennedy), and this raises the difficult question, can there be a stable undemocratic country in the modern world? Should the stable countries offer support to unstable countries in the hope of having them improve and stabilize, or will the support sustain an unstable regime indefinitely? Obviously, the solution is quid pro quo: the "no strings attached" philosophy of foreign aid namely the unconditional and unsupervised aid was based on a false theory, and it did intervene in the local political affairs of the countries so aided, as Sir Arthur Lewis has forcefully argued, precisely by supporting rotten and outdated power elites which would have otherwise been overthrown and which acted as cancers on the systems they controlled.

A simple example from international relations may illustrate the limitations imposed by the "no strings attached" policy. John Foster Dulles, Secretary of State under President Eisenhower, reacted to Egypt's flirt with Soviet Russia by a heavy-handed pressure which sent Egypt straight to

Russia's hands, and to the intensification of both internal and external problems. Foreign aid to Egypt, in the form of food for the starving masses, would have been far better politically, quite apart from humanitarian considerations. The "no strings attached" clause made it impossible for the U.S. to support Egypt while Egypt was flirting with Russia before the Cold War was quite over. But many alternatives to this policy were obviously open, not to mention the ease with which the Cold War could be avoided. This is not crying over spilled milk, but an uncontroversial example which can find contemporary parallels.

All this hardly scratches the surface. Even under conditions of peace and no fear of nuclear arms poliferation, unless population and the environment are controlled there is not much hope. These goals can hardly be achieved locally. One cannot expect that all who are willing to practice population control and family planning will cooperate unless some international guarantees to prevent those unwilling to participate increase fast both absolutely and relatively, so that Catholics, for example, will be quickly the vast majority of the earth's population. Since this is unacceptable to non-Catholics, it will not work. The same holds not only for religion, but also for nationality, cultural traits, and even such an insignificant matter as skin color. Of course, inasmuch as there is truth to the classical Enlightenment philosophy, parts of the world's population which will improve both their cultural and their economic positions will practice family planning and thus also population control, quite independently of the problem on the global scale. But even that will be the case only up to a point, and anyway the problem will not thereby gain any solution since rapid population growth prevents economic and cultural growth, as everybody knows by now.

Similar arguments hold for pollution, except that the major source of pollution are the industrial countries. As long as the world is afraid of war, overpopulated, devoid of stability and of global planning, the situation will not be under control.

The first step must, then, be true de-escalation of world tension. We know how de-escalation of tensions should proceed, but we cannot effect it without improving the situation in poor countries. Hence, it is much more useful, militarily, to aid other nations than to threaten them – which is the argument that has led to the vast foreign aid programs of the United States in the fifties and sixties. The argument is still valid – the programs failed because of the "no strings attached" clause, the covert pulling of strings (including political assassinations), and more. There are signs that this attitude may be changed, but only after Western countries will decide on a plan to effect change. In particular, it must be realized that it is useless to demand of the countries aided that they follow the models of Western democracies. It is much wiser to aid education and welfare programs than to demand hollow democratic legislation. Democratic legislation is pointless without some grass-root democratization, without some education and training for democracy (Habib Bourgiba).

The trouble is that we really have one theory of democratization – through the improvement of standard of living – which theory is empirically refuted. It is not only refuted by instances of rich and undemocratic societies like Kuwait and Bahrain. It is refuted by the fact that we can raise standards of living through advanced technology, and yet advanced technology may lead to technocracy, not to democracy (James Burnham).

Thus, there is a flaw in the idea that we'd better not suspect our neighbors but rather raise our standard of living and, by foreign aid, raise that of our neighbors: it leads to the tyranny of experts. The experts, to return to Russell's nightmare, will be glad to cooperate and keep the world dependent on them by keeping up armaments and suspicion of neighbors. The experts may allow the environment to deteriorate until they choke to death. But when they do die it may all be far too late. It is, therefore, particularly pressing a problem to see why technology, which is rooted in democracy, drags a technologically sophisticated system away from democracy and towards technocracy.

It is not hard to see where the trouble lies, even if it is not easy to pinpoint it. The force of democracy is in the diffusion of power through political institutions of decentralization, of control, and of balances of powers (i.e., of pitching powers against one another, preventing their unification). The force of technology tends to unite, as noted by Karl Marx, by Bertrand Russell and by C. Wright Mills, due to the absence of effective controls. The anti-trust laws in the United States are insufficient; in other countries they are absent; it is the multinational concerns that now increasingly control the markets. But why have controls over technological concentration not developed? The answer lies in common attitudes towards the law, in the disrespect for the law. And since international relations must be governed by international law, respect for international law must develop. And it cannot as long as the law of the land is not respected.

Respect for the law and the history of attitudes towards it calls for a separate study – especially the impact on it from recent trials of war criminals and of the civil rights movement. A study is also called for, of *Realpolitik*, which is the ideology of the license to politicians to lie. It gave birth to the constitution of the Soviet Union, which seems more liberal than any Western constitution, and yet justified the Gulag Archipelago. The starting point of reviving respect for the law should be laws violated in democratic countries by sizeable portions of the populations. Every educated citizen of the United States knows that prohibition was a disaster because people drank, were challenged and dared to break the law and get drunk, supported organized crime that supplied the booze and corrupted the law enforcement agencies, and demoralized the country. Yet they support laws against gambling, drugs, prostitution, speeding on the highway, etc. The present study is not the place to discuss such matters, but here is a proposal in quite general terms: every law must be enforced or abolished. The damage of the present situation is enormous, especially – to echo Spinoza – in the educational sphere.

5. CONCLUDING REMARKS

The link between the medium-range goal – the semi-utopia – and the immediate task, what we can do right now, the short-range goal, seems now to come slowly into focus: we do not go for the large targets directly, but through the processes of education – for responsibility no less than for knowledge and skills – and of democratization. And we need not fear that one who takes the lead with useful proposals and techniques may become the dictator or the technocrat if we develop our sense of responsibility and repeatedly give it institutional expression, in politics, in social custom, and in the arts. The question, what expression, exactly, need not be answered in advance and may be given more than one answer when the time comes, so as to have it debated. But we need to know more about the process of democratization and education, especially for responsibility, and about the institutionalization of results. For time is frightfully short, if it is not already too late.

CHAPTER 12

TECHNIQUES OF RAPID DEMOCRATIZATION

This concluding chapter has a practical problem as its topic: how can we prevent the technological apocalypse, the cataclysm forecast by ecologists, demographers, and political analysts? No solution will be offered here; only a very preliminary discussion towards an outline of a possible solution.

The main difficulty in plotting a solution to a mounting problem is to realize that when the solution will be ready the problem will have radically altered. Noticing this difficulty has led some thinkers to a most radical presentation of the problem: do we need technology at all? Do we need even economic growth? Can we limit growth? Can we block the growth of science? Now the growth of scientific knowledge needs no defenders, and its opponents should be dismissed as sheer obscurantists. Technology as such is, of course, unavoidable, only its level of sophistication can be increased and it can be decreased in almost every existing sector. What remains of the radical approach is at most the demand that technology or the economy cease to grow. This is to demand the impossible. It is also to demand the unreasonable: we must develop some sectors of the world's economy – at least as long as starvation is so very common. And for this we must develop some sectors of our technology, and we must curb other sectors. But even for curbing any sector – of the economy or of technology – we need develop new sectors. Our social machinery needs both better accelerators and better breaks and we must develop these and be quick about it. We must, in particular, develop the technology of controls.

1. SOCIOTECHNICS AND TECHNOETHICS

The most significant difference between Karl Marx and his latter-day followers in the Western world is in attitudes to public action. Marx thought one can start work for the world revolution immediately, and by helping simple workers attempts to achieve whatever they, the simple workers, are willing to fight for. Workers are willing, at most, to fight for better wages and better work conditions, said Marx. And they would be contented, perhaps, if they got them, though this is doubtful: the hovel grows smaller by the very fact that the palace next to it grows larger, he said. But, anyway, this matter is academic: under capitalism workers cannot achieve even the very humble and reasonable improvements for which they are ready to fight. The avant-garde, the intellectuals working for the world revolution, should not dismiss this good will. Rather they should help the workers organize for their own humble ends. Organizing, fighting, and having no success, the workers are better prepared to raise their stakes and even to hear about the world revolution and then even to join it.

So much for Karl Marx. His latter-day followers disdain helping simple workers to organize, since workers only aim at humble improvements: all they care for is more money and better working conditions; and they get these, and become defenders of the current regime instead of working for the world revolution. This thesis, anti-Marxist as it surely is, may be true or false – it is hard to say. Vladimir Illich Lenin endorsed it in 1905 (*What Is To Be Done?*) and concluded that he should make the revolution rather than lead workers in their fight for their humble ends – the improved wages and working conditions. Marx's latter-day followers outside the countries whose state philosophy is Marxian pray for the world revolution and fortify their faith in it by enumerating the faults in the existing system, by showing their contempt for it, by intensifying their sense of alienation from it. They are by and large politically inactive – although their criticism is often valid and at times does help others try to improve matters, and even though from time to time they themselves are willing to join or even organize mass movements with blessings from Lenin's heirs. One might think that political helplessness pushes the inactive out of the political arena altogether so that he can be ignored by the political activist. But this is an error: the helpless is harmful as he may occupy a position of responsibility which he thereby paralyzes. Also, the helpless justifies his inaction by an excuse: the action available is useless as it does not lead to the world revolution: the excuse discourages others. Quite generally, helplessness is contagious because the helpless develops a need, borne of his pain and frustration and self-contempt, to prove others no better than himself. This need, or rather this hope at a consolation by causing other people to be helpless, or rather the attempt to prove others just as unable to act as one's own self, is what makes the (politically) inept a sticky public enemy.

Marx was in error about facts, but right about policy. Workers do succeed in their fight for better wages and for better work conditions. However, they can use help and guidance in raising their stakes, and only one who joins them in their struggle as they comprehend it can help them raise their stakes.

More than ever now we have to fight for some kind of world revolution, though it should not be as naive as the socialist utopia envisaged by Karl Marx and his early followers. When the good people who followed Marx found that, contrary to his predictions, workers do succeed in their fight for better wages and for better work conditions, they got tragically confused and did not see how to raise the stakes of the fight and did not see what political end the fight might well serve, much less how it could integrate in a struggle for a world revolution.

What we need is still a strategy like Marx's: (a) a long-range – or rather a medium-range – alternative image of the world revolution; (b) a short-range alternative mode for helping simple workers – or rather any large portions of the population – to raise the stakes in their struggle for improvements; and (c) a strategy to link the two.

In particular, we should hope that under (c), not failure but success should raise the consciousness of the workers and prepare them to take the next step on the road. This is the vital correction to Marx's teaching.

All this is already happening, and in the movement which is already gaining momentum. It is variously known as sociotechnics or the humanization of work or for the improvement of the quality of working life or the movement for industrial democracy. The idea is that workers should fight not only for higher wages and better work conditions (hygiene) but also for ego growth through work (Abraham Maslow), and for job enrichment (Arthur Herzberg). This can be achieved by breaking away from the rigid hierarchy of work organization (Einar Thorsrud) and organizing autonomous work groups which take care of their own organization, arrange for rotation so as to have opportunities to learn on the job, to produce meaningful products and services, and to see to the quality control of their own product. It has been shown empirically that even on the lowest level of employment, workers do care about the meaning and content of their job (Judith Buber Agassi), so that organizing the workers for improvement can start at once; and it has been shown that considerable parts of management share this interest with the workers and can be mobilized to work for the same cause.

When the movement is really big it will be able to integrate work place and college; it will be able to break down ever so many barriers which were created or enhanced by the industrial revolution. Hence we have here a trend which may grow both on the local level and in political aspirations towards a better world.

Yet, clearly, the whole movement discussed thus far is confined to the workplace and is indifferent to a number of very significant global questions which are very urgent, such as the question, should the economy as a whole expand and in what direction, and how should the ecosystem be protected during the process? Here the goal is less difficult to discuss: we do not mind so much what exactly the economy will produce in the near or far future as long as the balance of the ecosystem is preserved. But we neither know how this is to be effected, nor how to tie it to any specific current political activity, group, or local interest.

Towards this goal the idea of technoethics has been proposed (Mario Bunge). Since modern technology was boosted by the erroneous idea that every increase of human ability to produce is to the good, and to the extent that the consequence of this idea today is a danger to the ecosystem as a whole, we need a new moral code for technological daily activities to guide both researchers and industrial organizers, if not every member of the community. The ecological movement did, indeed, begin some work in this direction, advising people to purchase food not grown with the aid of fertilizers, not to purchase bottles produced for single use rather than for reuse, to separate different items of refuse so as to raise the level of recycling. The governments of a few Western countries, especially the federal government of the United States, began a campaign to save energy, legally reduced speed limits and levels of heating of homes in an effort to save fuel, and more. All these activities have thus far proved useful in one and only one respect: the general public is consequently better aware of the problem. But there is no solution in sight and no policy which might evolve into a trend leading towards any reasonable solution. In particular, to eat natural food, i.e., food grown with the aid of no chemical fertilizers and prepared with no additives, does not solve any problem of the soil and does not save the consumer from pollution as long as he drinks polluted water and breathes polluted air. Indeed, the personal standards in question are the fairly traditional norms against littering extended a bit to consider certain acts of pollution as littering even though the litter does not show. This is fine, but it will not do.

There is an obvious limitation to what can be achieved by personal ethics, even on the level of rules against littering, or against jumping the line, or even against petty theft: the rules can be imposed only if the majority of the population find it not too hard to keep them on a voluntary basis, make them into habits, and incorporate them into the national style. Thus, in a stable civilized country, if one person or two jump the line, the rest may tolerate the fact or they may fight it, but they will stay in line; in many countries one can see a line with people standing patiently, and then the line dissolves in a few seconds due to one person's impatience and misconduct. No doubt, this makes the introduction of new norms particularly hard, since the public must be assured that people who endorse the new norms will not remain in a small minority for long. In the West the standards of no littering have risen and broadened to include some ecological imperatives, but as long as the military and industry pollute and waste on a large scale it is hopeless to expect the private citizen to do much better. The private citizen will not stop littering and save fuel when one war plane on one training mission pollutes and wastes more fuel than ever so many automobiles do over a whole month.

That personal ethics depend on the moral standards of organizations – in the private and in the public sector – is not new. It is similarly acknowledged that business and government ethics have to grow together with the individual's personal ethics. And the problem is that the standards differ: for a person to take bribes in the course of his public duty may be very similar to petty theft; but for a corporation to bribe a government official – at home or abroad – is immoral in a different way: unless it is made imperative by legislation or company policy the honest employee will be harmed by the more unscrupulous, loyal, ambitious one. Hence, it is incumbent on anyone concerned with ethical problems whose solution is blocked by organizational behavior, to participate in processes of public decision-making – company policies, legislation, and more. The attempt to boycott products of companies or countries violating an urgently needed code is of this kind but is usually useless. This is not to say that all boycotts are equally useless. Mario Bunge has suggested we begin by advocating the boycott of research into evil practices in the hope that it may develop into laws or company policies to avoid evil technological research.

We can aim at integrating sociotechnics and technoethics. Employees with more understanding of, and more control over, their own work tend also to demand that their company avoid polluting the environment or otherwise behave illegally and/or immorally. The control of various ethical aspects of a university is usually done by a democratically elected committee. The more autonomy and democracy enters the large workplace the more its employees could participate in monitoring ill effects of their products or services on consumers, community and environment.

All this requires much further study, careful planning, experiments, pilot programs and public discussion. It is a mere hint as to the possibility of connecting short-range and longer-range goals.

2. TOWARDS A THEORY OF RESEARCH INCENTIVES
AND ASSESSMENT

Research incentives and assessment are these days very central in debates in the field known as science policy – debates about both technological research and technology assessment. The reason is that technology is often viewed as sophisticated and science-based. The most expensive large-scale experiments are taken as the model – such as the making of the bomb and subsequent growth of nuclear energy plants, and such as the space program and the subsequent development of the microcomputer. Hence lumping science and technology together seems amply justified.

This is rather lopsided. Yet it is understandable: we may reasonably suppose that if we characterize big research we also thereby characterize to some extent all research. We may see this when we examine the current popular debate about the question, how much modern physical and biological technology proper owes to modern science proper? Since this debate takes the difference between the two as a matter of course. It should be noted that the previous paragraphs present the fact that technology is usually lumped together with science and this paragraph refers to a popular debate based on their distinction. This is hardly surprising: any two things may be lumped together in one discussion and distinguished in another. What is needed is to notice this fact and maintain a sense of proportion.

The real question is not so much as to how much science contributed to modern technology in the sense that scientific ideas were applied or not applied by this or that technologist. It is, how much of the technological style (in the sense of Lynn White and Heather Lechtman) of the industrial revolution is carried over from the scientific style of the scientific revolution? The very question forces one to see the debt of the industrial revolution to the technological revolution and the debt of the technological revolution to the scientific revolution. The stupendous developments in the nineteenth century, which are processes of industrialization of a large scale yet owe little to technology and less to science, need not make one conclude that it is divorced from scientific influence. The ideas which revolutionized industry and agriculture were at times purely organizational, at times simple technical ideas with immense organizational corollaries, at times they concerned the use of advanced science – but they always reflected the Western scientific ethos. Many conspicuous advances are due to advanced natural science rather than to purely organizational ideas, like the invention of the small share or of replaceable parts, or of simple technical ideas.

Hence, the choice of nuclear weapons and space program and similar "big science" items do, indeed, capture the scientific style that characterizes much science and technology; the lopsided picture presented to the public is reasonable. This is not to say that the picture is not lopsided. It is even dangerous. Though the work process in the chemical industry owes much to the style of science, it is still divided into routine jobs. The routine job is the opposite of research and needs redesign. Also the small research project fundamentally differs from the large one. On the one hand, it may be conducted without financial incentives; on the other hand, it is easier to get support of a small fund from a grant-funding organization than a large one – for purely political reasons. One way or another, the trouble with most researches, including most small independent ones, is that due to our lopsided picture of science and technology all researches are infected with the style of the big ones – the enlarged style of classical science. The pioneer of enlarged science was the inventor of the first enlarged incentive – Alfred Nobel. He noticed the inadequacy of the classic style when he saw his invention (dynamite) put to a large-scale military use; and he hoped that large-scale incentives will save small independent research.

The fact that we scarcely refer to social technology is also due to the seeming backwardness of the social sciences and the doubts as to their profitable applicability. These are made conspicuous by two comparisons. Let us compare, first, the work of a twentieth century scientifically trained chemist who is not in the least a scientist, whether employed in a drugstore or in a chemical plant, with any chemist employed anywhere a few centuries ago – whether an apothecary or a mining engineer or a manufacturer of any chemical whatsoever. The distance is enormous. Second, let us make a similar comparison in any social technology other than in finance economics: it seems that no administrator today uses sophisticated techniques at all comparable to those of the chemist. The introduction of the computer into business (and even to voting estimates) refer more to physical matters like taking stock and ordering commodities and calculating costs. These, however, seem not related to innovative social organizational affairs, hardly related to creating new ways of handling social problems. Some social problems have vanished (we are no longer worried about duelling or lynching), but hardly due to the application of social scientific ideas or even the style of science in any way. The applicable social scientific theory is the view that the penal code as such is foolish – an idea which became popular partly due to the utilitarianism of the style of science, but it is an idea known to ancient sages and religious leaders alike. Nevertheless even this old idea is not yet as widely applied as possible, though it did bring about some important reforms.

In brief, the impression which is widespread today is that no social research has led to important practical results.

Any person concerned can attempt to contribute to research, and in diverse ways. We need, in particular, research into current research practices, criticism of them, and proposals to reform them. We can begin by the study of the views current today among leading research grant givers, especially the military; among leading researchers, especially organizers in the guise of scientists; among leading prize givers, especially the Nobel Committee; and among technical advisers to governments and to large concerns, especially to the leading universities. It would benefit the world if one could make public a critical assessment of their views about the way proper research should be planned, written up, evaluated, executed, reassessed, and prolonged or terminated. This might open the road to new sets of alternatives to them which one might discuss critically.

These matters may demand hard work from interdisciplinary teams. But the assessments of simple assessments of projects by a given organization over a span of a few years can be performed with ease, and might easily expose the sabotage that teams working on expensive projects exercise on critics and on small-scale competitors. (Large-scale competitors merge, coordinate, compromise.) The techniques of sabotage are fairly simple and easy and so are the counter-techniques of blocking the sabotage and keeping the freedom of speech and of research. Also, the sabotage is usually meant to block things not for good, but only for the duration of the costly research. Hence the counter-techniques should cancel the time lag.

Some criticism is systematically suppressed – for example, the criticism of the medieval guilds in modern societies, such as the medical, legal, and accounting professions, as well as the whole university system. The guilds are, first and foremost, self-serving closed clubs. They are pernicious in suppressing much research or its fruits – whether about Jenner's vaccination or about the value of the accessibility of medical records to patients. Useful research might aim at dismantling the guild, i.e., opening the guild to public democratic control.

The threat of (temporary) suppression exists in physical technology as well. Whole industries are ousted when technology creates substitutes for them. Leaders of industrial concerns know this and so may be willing to suppress research. And they may choose to suppress only implementation of innovations, not the research leading to them – and the suppression may be effected by the raising of the standards of test required by law or custom – whether the test of the effectiveness of the innovation or of the absence of side effects. Or, as Lenin said, a concern may acquire a patent in order to suppress the patented invention. But usually in the Western world industry does not suppress physical technology; rather, it proliferates and tries to have a part in the new fruits of research. For example, new electric watches ousted much of the market for mechanical watches, but the large watch companies simply produce both kinds – indeed many kinds: industrial concerns learn to adjust. When a firm is caught unadjusted it may go under, but usually it is purchased by other firms and is then helped to adjust. When a whole national industry is unadjusted, this may cause a real problem – such was the case with the Indian indigo industry when artificial dyes were invented in the West in the mid-nineteenth century. In the developed world, when a whole national industry is caught unadjusted, other industries or the government have to step in – as was the case with the American auto industry which refused to take account of change in aggregate taste – the preference of small, fuel-saving cars. This failure was not due to weakness of the physical technology – on the contrary, Detroit was more used to changing its models than the (more advanced) European and Japanese industries – but due to poor social technology. The defect, to be more specific, was not so much due to inability but due to the classical scientific style – due to the lack of incentive that is conducive to irresponsibility. All the Detroit companies gambled together, in the knowledge that the nation will not be able to let the whole of the American auto industry go down together. The political-economic leadership, then, could have easily foreseen the hardship of tens of thousands of laid-off auto workers and their families and the cost to communities and to the taxpayer, since similar experiences occurred elsewhere – even in the auto industry (in Britain). Research could be designed into such acts of lack of national responsibility and into the needed appropriate safeguards.

What is important here is the ability to empathize with people who feel threatened by social research and who feel justified in subsequently meeting what they see as a threat, thereby risking lives, even life on earth as a whole. What is called for is the removal of the threat – the institutional reform that will remove what makes these people feel threatened so as to make them feel safe. This can be done relatively easily in democratic countries. How to do so in undemocratic and/or backward traditional society is much harder (see *The Destruction of Nature in the Soviet Union*).

But the hardest is to have an open democratic organization that will offer incentives and exchanges of information concerning research about re-search. If at all successful, such an organization might become enormously powerful and invite the rabble and the mixed multitude to jump on its bandwagon. This is a general problem for all democracy. Perhaps it should never be solved; perhaps what keeps such an organization young is its ability to transform selfish new recruits into enthusiasts – into active students who bring with them new approaches, new ideas, new zest. When they grow old, new young ones have to be designed (see Russell's *Zahatopolk*). In this, too, as in everything else concerned with the future of mankind, research into research should be largely educational. Since education, however good, relies much on the public's educational background, a major area of research should be the evil of what goes these days under the label of educational research; in particular, we might study the risk of mounting our powerful modern computer technology on our backward-looking misanthropic edu-cation. This can be found in the leading high-powered institutions for higher education, which are aimless tough-and-no-nonsense establishmentarian bodies, training apprentices to hunt for all the rewards and accolades of their profession and of their society, without ever stopping to ask what values these rewards and accolades represent. The universities always resisted social reform and moral rejuvenation and may be trusted to do so especially vigorously regarding efforts to democratize them. As David Budworth has noticed in his autobiography, it is hopeless to attempt to convince (British) professors that industrial research is a worthy activity, the training for which does not necessarily follow the lines of training they envisage: the professors take it for granted that university style research is best and they are not obliged to listen to any outsider. The only way to render their stubbornness on this matter harmless to others, he observes, is to break the monopolies universities have on our educational system. Albert Einstein had a simpler plan. He said, it is very easy indeed to reform the educational system: all one needs is to take away the power of the teacher. This I think should be the guideline of all educational reform. The transfer of the power to give licences to practice – to lawyers, doctors, teachers, engineers – from the university to some democratically elected and democratically controlled bodies should suffice. The bodies may be public or governmental. They may be run by professionals and experts or by professionals and amateurs together. They will be independent of the university and this will be a great relief.

3. TOWARDS A THEORY OF DEMOCRATIC MASS-MOVEMENTS

Classical philosophy made no provision for the adaptation of society to technology, no provision for the social reforms necessitated by technology. Though social changes of this sort were made, they lagged behind and were implemented ad hoc. This will not do now: we have urgent questions on the agenda, concerning social changes due to technological developments – what social changes ought we introduce, and how can we introduce them rapidly so as to avert too much of a calamity? To narrow down the question, so as to render a preliminary discussion of it possible, let us see what we can learn from the recent mass movements about rapid social change. Can we make the mass movements more effective, more democratic, more instructive? More pointedly, can we focus the mass movements on the technological apocalypse?

The problem is already adumbrated in the autobiography of Bertrand Russell, the father of the recent mass movements, or at least a major factor in their evolution. Russell did not plan things in any manner that resembled the outcome. He had an immense sense of urgency, a sense of now-or-never about the choice between abolishing nuclear war and abolishing mankind. He felt that the choice was in the hands of the fates, whereas it should be made rationally by all concerned. Today we are prone to forget this because his Ban the Bomb movement ended in a failure of sorts, and because somehow, perhaps miraculously, a precarious balance is kept and we pretend to have learned to live with the bomb.

Russell was politically active all his life, usually by lecturing, writing, explaining, teaching the public as best he could. In 1957 he wrote his exciting open letter to the leaders of both the United States and the Soviet Union, which opened up a sort of public debate between them. The debate dried up fast: the Russian public could not participate and the American public was not sufficiently interested. Russell gathered a few like-minded people who likewise tried to arouse public interest, and together they founded, early in 1958, the Campaign for Nuclear Disarmament, CND for short. The CND organized the celebrated Aldermaston Marches in 1959 and 1960, the mass meetings in Trafalgar Square in the heart of London, and more. "By the summer of 1960 it seemed to me," comments Russell, "that Pugwash and CND and the other methods that we have tried of informing the public had reached the limit of their effectiveness. It might be possible to so move the general public that it would demand *en masse*, and therefore irresistibly, the remaking of present governmental policies, here in Britain first and then elsewhere in the world." Russell knew what change he wanted, and he knew he needed mass public pressure in order to effect it. But he felt his techniques fell short. "Towards the end of July, 1960, I received my first visit from a young American called Ralph Schoenman ... I found him bursting with energy and teeming with ideas, and intelligent, if inexperienced and a little doctrinaire about politics ... What I came only gradually to appreciate ... was his difficulty in putting up with opposition, and his astonishingly complete, untouchable self-confidence." Here Russell describes Schoenman as the archetype of the mass movement organizer, who has since become a familiar part of our scene. "At the particular time of our first meetings," continues Russell, "he acted as a catalyst for my gropings as to what could be done to give our work in the CND [Campaign afor Nuclear Disarmament] new life. He was very keen to start a movement of civil disobedience that might grow into a mass movement ... so strong as to force its opinions upon the government directly. It was to be a *mass* movement, no matter from how small its beginnings. In this it was new, differing from the old Direct Action Committee's aspirations in that theirs were too often concerned with individual testimony by way of salving individual consciences."

Russell reports their subsequent activities cryptically. He does not analyze or explain, and his narrative from this point on lays stress on civil disobdedience – an indubitably new ingredient in the campaign, though in itself not new. The mass movement began early in 1958; Schoenman entered it in 1960; the activities in the new style burst upon the world in winter 1961 and the momentum picked up and stayed high for almost a whole incredible year – between fall 1961 and summer 1962. In January 1963 the whole affair was over. So much for Russell's story and his direct contribution to it.

What then happened has not yet been sufficiently chronicled, but is still fresh in memory. The movement crossed the ocean and spread in the United States in diverse directions: student liberation, black liberation, sexual liberation, women's liberation, gay liberation. But all these movements were, for most of the time, put in the shade by the mass protest against the American involvement in Vietnam – ever since the day Martin Luther King Jr. declared he could not go on in good conscience leading the black liberation movement without joining the anti-Vietnam war movement as well, and until the end of that war. The movements, especially the student movement and the anti-Vietnam war movement, spread all over the world. Their techniques included, as had the black liberation movement before, both civil disobedience and violence. What the students introduced first were the teach-ins. These were immensely popular and successful; nevertheless they were stopped, since they were viewed with suspicion as possible means of slowing down the movement and thus dampening its impetus and robbing it of its mass character.

Soon after the Vietnam war was over, much of the impetus dissipated. Some of it went into a new mass-movement – the ecology movement. This was severely curtailed by its own lack of direction and by the energy crisis. The other movements had some other measure of success and some consolidation. Large and powerful and beneficial as their results may be, they can hardly be viewed anymore as mass movements à la Russell and Schoenman, since either they consolidated and became organizations proper or their demands are accepted by existing institutions. The women's movement still has some characteristics of a mass movement, exhibited in some new organizational tools which are relatively informal and fluid yet institutional all the same.

There are successes and failures to be assessed in many respects. What concerns us here, however, is the moral from this for a possibility of more enlightened, even sophisticated, mass movements. For this democracy and rational exchange are of the essence. Is there no conflict between rational dialogue and mass movement?

Democracy, it may be argued, like any other political form, presupposes some stable organization, some purpose, whereas mass movements of all types, being movements, are not well defined for the question of democracy to hold. To this one can answer, first, that even the most amorphous body has some rules of procedure, however loose, and rules and procedures may be more democratic or less democratic. Second, the same holds for purposes.

Bertrand Russell conceded that there was room for debates, that a mass movement suppresses debate, and that hence his support for a mass movement was a matter of expediency. In his autobiography he says nothing about the failure of his movement, partly in order to avoid squabble but partly also due to this difficulty. The leaders of the mass movement in the United States were in much less trouble. Professor Noam Chomsky declared, "I belong to the party that says that the grass is green." During the days of peak activities, Boston University students stood on top of cars at the edge of Boston Common under the glare of the television cameras and yelled through bull horns to thousands of their peers who filled the common. We tried all possible peaceful means, they declared earnestly; all means failed: it is now time for the revolution. This simplistic view of things stems from the line of thought of Marx and Lenin, and was conveyed to these youths in the successful classes of Professor Howard Zinn. But that is not how things work. The charming, earnest, sincere youths had not tried all the peaceful means.

The idea of a mass movement then was really simple. The issue must be clear, the awareness of it wide enough, governments have to be as stubborn as the Pharaoh, and all the organizer has to do is call all the Children of Israel to gather together and force their way on one concrete issue. They can force governments to pass certain laws, or populations to give up certain bigoted attitudes. And the change has to be such that reversal is practically out of the question. This may indicate that through mass movements we can achieve only what is adequate for the intellectual level of the population. But there are different components to the education and sophistication of any given population, and, moreover, such things can be developed, and in mass movements they can be developed rapidly. On this point, on the possibility of rapid mass education, Marx was right. It is possible, and, as he said, it can be achieved only through combining the struggle with the education of the struggling masses. He would have greatly approved of the teach-ins, which were immensely successful but were stopped, first by organized heckling by the Students for a Democratic Society or the extremist branch of that movement known as Weathermen, and then by the organizers of the movements.

Who organizes mass movements? Who has the right to do so? When?

Perhaps the most wonderful thing about mass movements is that anyone who thinks he is able enough and has the right conditions can try. It is cheap to dismiss mass movement organizers as charismatic. The immensely popular screen actors Paul Newman and Marlon Brando joined mass movements, actively, but they were not the leaders that Professors Chomsky and Zinn were. If the very success of a mass leader proves him charismatic then that is that. Yet democracy is best served by allowing any one to act as a mass leader. Even the fact that the major tool of some mass movement is civil disobedience does not change this. And of violence and inciting to violence, suffice it to notice that the bitter experience of the American Black Panthers proved by tragic means the obvious truth that violence can serve a mass movement only in the popular overthrow of a tyranny.

The movement that has the greatest promise for technological problems and that should undertake the greatest and most important and urgent roles is the ecological movement. That movement developed rapidly – as rapidly as other movements – partly because a vacuum was there to be filled in the space of mass movements (the vacuum is still there), partly because of the new and intolerable level of pollution (the situation is rapidly deteriorating). The U.S. movement was defeated – as a mass movement, that is – by its inadequacy. It chose three main issues to fight about: the Alaska oil pipeline, supersonic civil flight, and the pollution of Lake Erie. It claimed that the Alaska pipeline issue was clear enough and won, but the energy crisis immediately reversed the decision – to the satisfaction of many ecologists, as it happens. It declared the supersonic commercial flight clearly unacceptable and won there too, but technology improved the level of performance of supersonic flight and thus proved the issue not at all clear-cut, not irreversible, nor really related to the general urgency of the situation. Mass protest about Lake Erie did not help decide how to save it. On the whole, nonetheless, the movement did much to bring public awareness and make ecology a profitable political plank everywhere. In this way the movement did help save the lake.

But this is not the end of mass movements: the modern-style mass movements simply sprang into being under stress; it is of their nature that they come and go as they do, with intense stress and while heightening the excitements. Hence they must remain to a large extent spontaneous. Of demonstrations the last word still is Lenin's. Demonstrations, he said, are hard to organize; spontaneous demonstrations, he added, are particularly hard. Mass movements are prolonged spontaneous demonstrations and so very hard to organize; democratic mass movements are particularly very hard to organize.

The interest in mass movements, then, will keep them going and lessons from them will be seriously considered by the able organizers of future ones. It is therefore easy to notice that as long as the West, where mass movements succeed which manage to present a clear-cut issue and effect those changes, in public morality, in administrative or court procedure, or in the law, which take root at once in the society in which it is effected. (This is not the place to discuss mass movements in communist and other countries, but certainly the Polish and the Iranian experiences should qualify as involving mass movements, however different these are from the ones in the democratic industrial world.)

It seems that the future of mass movements is ripe for two major changes. First, that its organizers will have to present both a clear-cut case, and also a clear-cut argument to explain why the expected change will take place and not be reversed after spirits quiet down. Second, that the major tool for forging clear-cut cases can be the teach-ins plus the mass media. There is everything to say against masses acting administratively, since government by the people is impossible. Nor should masses act judicially, since mass tribunals are barbaric. But there is everything to say for the teach-in mass demonstration to emulate the practices of parliamentary and congressional committees which can invite expert witnesses and interrogate volunteers. Here lies the future of this medium of social and political change. For this the organizers will have to start their procedures a bit ahead of time so as to be in phase with the turns of events that prescribe rapid action. For, the main rationale for this medium is that time is short.

This brings us, then, to a philosophical aspect of the matter. The problem of induction as a problem of empirical justification of action, social or private, is insoluble. We never know whether we are too slow or too fast in implementing an innovation. Different societies have standards regulating all this, and the standards are regularly tested and altered. But some innovations are not subject to standards, some standards vary greatly depending on the urgency of the situations. Military establishments take greater risks in testing and implementing innovations since they fear the greater risk of unpreparedness; market mechanisms push corporations to similar considerations. Pilots say runways only improve after blood is spilled on them; because, presumably, runways conform to standards but standards are inadequate and improve too slowly.

That population control and pollution controls are matters of emergency is commonly admitted. That standards to deal with them are either grossly inadequate or nonexistent is likewise admitted. The mass movement can come in here. It will make mistakes like any other movement. This should be no discouragement if it is readily admitted, especially since the mass movement, being so spontaneous and almost entirely amorphous, can be more flexible than any organized body – on the condition that we can clearly face the emergency without falling into hysteria: can we present reasonably the dangers of the oncoming technological apocalypse? Can we offer a sufficiently simple and obvious agenda on it for a mass movement to work on?

To conclude, since mass movements grow spontaneously, they must be planned and organized around clear-cut issues, on which legislation is more likely to be irreversible, and act mainly educationally to create a new awareness in vast populations through discussions on proposed solutions that seem sufficiently clear-cut and irreversible. We do not, however, have solutions to our central problems. We may start, then, with the idea of altering our standards of political and diplomatic truthfulness and demand the use of this as some sorts of means for enhancing the value of international bodies which enhance peace. This proposal is far from being satisfactory. Whether some significant additions to it will be made, can hardly be foretold. But since global coordination is essential today and since the West did not prevent the Soviet bloc and the oil-producing countries from destroying what little value the United Nations Organization had, we may hope for mass movements forcing the governments of the leading nations to keep negotiation going on and report truthfully about it. And the same should hold concerning disarmament talks: they should be as open and frank as possible. This way we may hope to create new and more truthful and more effective coordination for global emergencies then the United Nations Organization has been thus far. The need to rehabilitate this body, and by the imposition of truthfulness in its meetings, is all too obvious.

4. TOWARDS A THEORY OF RESPONSIBLE DEMOCRATIC LEADERSHIP

Classical liberal philosophy declared all individuals autonomous and so in principle in no need of leadership or of laws. The Romantic Reaction declared the multitude heteronomous and in need of leadership and the leaders autonomous – heroes whose words are laws. The theory of autonomous leaders of autonomous law-abiding people – the theory of democracy – is still absent; perhaps the theory should handle the populations of modern democratic societies as at times autonomous at times not, at times responsible at times not, at times competent, imaginative, resourceful.

Because traditional philosophy sought the justification of power, it justified either only ideal states or all of them, including the worst that exists. There were exceptions, to be sure, such as the philosophy of Spinoza. He too thought autonomous people need no leadership and no laws to guide their conduct, yet he thought democratic leaders and laws are needed as means of educating the multitude so as to make them autonomous. Even this is too impractical and besides the point.

That autonomy and leadership look contradictory is too obvious: the autonomous decides for himself and so needs no guidance – of leaders or of any one else. Yet autonomous people freely follow their travel guides. The travel guide is the agent of the person he guides; the guide is responsible to the guided, and can only offer his guidance, not impose it. The guided who is obliged – for any reason at all – to follow the instructions of his guide, being obliged, has lost some (or all) of his autonomy.

All this is not yet a critique of classical individualist political philosophy, although one should develop it so as to show that it is such a critique and that therefore classical individualist political philosophy has been superseded. To begin with, classical individualists had no doubt about society's need for coordination and about coordination having to be effected by political means so that to meet that need the institution of political organizations is necessary. The classical discussion centered on political freedom, not on coordination, and thus it questioned political power, not political coordination. Somehow, the correlation of coordination with power was usually[14] left unexamined. The important question is not, is the politician's power justifiable, but, can there be coordination with no power? The example of the travel-guide is thus irrelevant unless he is responsible for a travel undertaking which invites coordination.

The fact that a travel-guide may wield some political power is both very well known and systematically overlooked. We all freely talk about national politics as well as local politics and distinguish these from the politics in the work organization, in schools, in clubs, in every social body. We even distinguish between the politics of a political party and the politics within it. All discussions of politics within organizations other than the broad political ones, whether politics in a work organization or in a university, whether politics within a national political party or within a local chapter, within the national trade union or in the local chapter – they usually have to do with the demand for democratization. The theory of participatory democracy which has burst upon the intellectual scene after World War II has developed into the claim that all politics should be democratized, where politics is the power that is required as means for the implementation of coordination of matters significant for the whole public in question.

The contrast between classical and modern individualist political philosophy begins thus to emerge. To begin with, classical theoreticians discussed only the power of national central authorities. They may have felt amply justified in ignoring smaller and not quite political authorities such as the power of the employer over his employee. There were strong reasons for this: national central power was more problematic; it also could control other powers; and employees can more easily switch employers than nationality. To this day economists who support classical theory on account of its individualism and liberalism oppose democracy in the work place even though they know very well that the liberty of a worker is unnecessarily constrained by excessive employer power over him.

Secondly, it becomes clear that the question of justification of political power has to center on the question, how much power do we need in order to efficiently coordinate certain bodies and certain activities. We can thus consider how much police power has to increase in order to reduce crime and then seek the optimal condition between full anarchy and total dictatorship; also, as usual in the theory of choice, there are alternatives to police power, such as legislation aimed at the creation of incentives to be law-abiding, such as education, such as the development of good relations between the local police and the local population, especially its poorer members. This last point, incidentally, strongly relates to the topic of participatory democracy: the more the population participate in the democratic process the better: participation teaches people the rules and the feel of the democratic process and raises their level of involvement in politics – both actual and potential.

Third, the classical theory deals with the actual power of any potentially justifiable government, and concludes that such a government should do almost no coordinating – the classical theory prefers coordination to be effected voluntarily by non-political bodies, leaving to the government the role of a night-watchman. Here the very idea that any coordination requires power is finally lost, since thereby power is given to the government which should not coordinate and coordination to voluntary organizations, especially economic and commercial, that should wield no power. But it is the effective ability to coordinate that requires power, and it is the very extent of the coordination that suggests the kind of politics at hand.

Much has been written on the question of centrality of power, and liberals and pluralists and particularly liberal pluralists opt for decentralization when at all possible. The question is, how does one decide such matters? Centralization increases the possibility of rapid and controlled coordination, and even inefficient, non-democratic, yet centralized governments can at times exhibit bursts of efficient coordination not even conceivable under democracy. Yet centralization is also possibly inefficient – depending on the control in the hands of the center – particularly in their being detrimental to proliferation and to experimentation. How does one decide such matters? What, to begin with, is the legitimate domain of a central political power? What should national central governments take as their proper responsibility?

Again, classical liberals, from Hobbes to Smith, said, the keeping of the peace is the sole function of the central government and it should do no more than necessary to keep the peace, taking the citizen's own personal interest to be politically neutral. This is why Hobbes, who recommended absolute dictatorship, is put with Smith in the liberal camp, whereas Thomas More, who thought each individual should partake in his community's affairs, is not. In a strange sense participatory democracy attempts to unite liberalism with populism by declaring that though the individual is free to be politically inactive, he should be free to be politically active, and this latter freedom should be catered for by the central government. This returns us to the question, what is the proper domain of a national central government?

This question must stay open and follow the decentralist rule: every question that can be handled locally reasonably well should be handled locally; the central government should see to it that this rule is properly implemented; the problem that calls for central handling becomes a national issue for as long as it is nationally pressing. And, of course, foreign affairs, defence, and a national currency must stay central. As long as the central government cannot be expected to handle defense and the economy without a centralized military establishment and effective treasury, these must remain centralized, but otherwise not. For example, in some countries a central police force seems essential, in others the police can be decentralized and often it is. The neo-classical economists cannot explain their demand for a national currency because the classical theory does not speak of the potential political needs; it speaks of the actual needs of the ideal government instead.

This idea, that governments may be organized with the idea of potential needs is particularly dangerous. We all agree that at times of emergency democracy may be suspended. Once we allow potential emergency as a standing factor, democracy is gone for good. Without some arrangements for potential emergency, government is irresponsible, yet a government constantly geared to emergency is undemocratic. Where is the reasonable dividing line?

The question has not been discussed by any philosopher. The justification of political power has led not to democracy but to the authority of science, whether elitist or populist. The definition of demo-cracy as government of the people led to the justification of any government elected by a democratic process, no matter how undemocratic; by the people led to confusion; for the people to benevolent tyranny. Sir Ernest Barker's definition of democracy as government by discussion is excellent, especially when understood – in retrospect – as participatory democracy; but it leaves too many questions unanswered. Sir Karl Popper has offered his definition of democracy as the government which can be overthrown by the people with no bloodshed and, as the regime where such overthrow is instituted – usually by general elections which are routinely compulsory. This definition is by now very popular. It seems clearly necessary but not sufficient. The claim that it is sufficient makes government minimal government, so that one can see Popper's theory as an improved variant of classical political philosophy. The reason is very simple. It is a theory which only attempts to limit government powers and which does not empower government to do anything in particular; it is not a theory demanding that government coordinate, control, or defend the nation. Moreover, taking this theory literally we can only blame a government for doing something unpopular, not for failing to take responsibility and propose to the nation new ways of handling problems. Furthermore, still taking the theory literally, there is no reason why government should not be elected for the shortest term possible, and replaced any time it attempts an unpopular measure. It is common knowledge that this brings about the instability of democracy which is the strongest and commonest cause bringing populations to favor a tyranny in preference to democracy. Clearly something has gone wrong here. Popper's theory of democracy, if implemented as it is with a vengeance, is dangerous to democracy.

The question, How stable should government be? is practical, not philosophical. Different democracies practice different procedures and try different ways of finding the happy balance between the positive and the negative. Hence, it is unreasonable to expect a general answer to the question – at most one can offer a general criterion. Yet, if we insist on Popper's view of defining democracy only negatively, only by demanding that government be not too stable, then the simplest way to satisfy this demand is, democracy had better be as unstable as possible. The fact that we all reject this answer, shows that we think democratic government, like all government, has a positive role to play. The enemies of democracy say, the positive role demands more stability than democracy allows. This is an error – a refuted prejudice and no more. But is it inadmissible to meet this prejudice as Popper does and say, philosophy should only care for the negative side: we have to take care of the positive side, of the government's positive responsibility.

Popper is still right to a large extent – democracy must have mechanisms for the peaceful change of government, and there is no need for a general theory to specify the mechanism. Yet democratic government must do more. What more it has to do can also be said not in detail, except that there must be a mechanism·for the peaceful change of the national agenda. Also that survival is the first item on every national agenda. How high peace is on the national agenda is debated among philosophers, from pacifists to utilitarians. In today's world the debate becomes increasingly academic since world peace becomes increasingly a matter of the survival of humanity. But the question of the mechanism of the construction and change of the national agenda calls for one or two additional comments.

The classical rationalists aspired to develop a political science free of error and thus develop the right agenda. The classical Romantics thought leaders knew intuitively what the right national agenda was. Nowadays we cannot hope for any insurance against error, but we still can hope to avoid irresponsible error. And to this end we can hope to develop a responsible national leadership.

Democracy was declared to be the government of the people by the people. This was never taken literally, yet it did leave a lacuna – we still have no theory of democratic leadership. And we need a theory of responsible democratic government. For, it is an empirical fact that responsible leadership can replace an irresponsible one and take over the government – and when possible without bloodshed, as the rules of democracy require. But when there is no responsible leadership to replace the irresponsible one which is in power, then neither democracy nor any other system will save the nation.

How, then, can we safeguard that government can be replaced by peaceful means as well as that the nation has responsible leadership both within and without the government? This is the new question which democratic political philosophy must face. The novelty of the question lies both in the new accent on responsibility in addition to autonomy (a point implicit in Popper's writings already) and in the claim that responsible leadership lies not only in the hands of the government but also in the hands of the opposition and of some civic bodies. This is political power at large – a political potential. It is regularly placed in prominent organizations outside parliament, such as the broad organizations of the parliamentary political parties, the voluntary political bodies, and the mass movements. How are these possible?

The very existence of political leadership outside government circles and outside the parliamentary system is an important fact. In part it must be so because politics is not limited to the national political system alone, especially in countries where some political leadership is regularly selected out of the leadership of economic, industrial, military and intellectual spheres. In part, however, it has to do with the combination of the talent and the sense of public responsibility of those who engage in politics on occasional basis only.

Responsibility is something which has entered political philosophy only after World War II. Before that there was more responsibility than today but less talk of it. The aristocracies of the different societies always took it as their responsibility to govern their people well. *Noblesse oblige*, they said. But this was increasingly an excuse for discrimination rather than an expression of the acceptance of an obligation. The result was that democratization did improve matters. And since the enlightened philosophers were defenders of democracy against aristocracy, they played down the controversial matter of responsibility altogether. Theoretically, they reduced it to autonomy: responsibility can only be to one's own self or to one's autonomous peers if they appoint one to a political position. When the reaction to this philosophy evolved, it declared the leader autonomous but his people heteronomous, and his autonomy then expressed his responsibility not to others but to himself and for his nation and/or History. This enabled utterly irresponsible people to take over and act recklessly while expressing a profound sense of responsibility – to their own selves and to History.

Responsible leadership can be seen in all societies, also where the autonomy of the individual is not preached. Admittedly, as anthropologists observe (Sir Edward Evans-Pritchard, Ernest Gellner), aristocrats are often autonomous by virtue of being educated to be rulers. Their autonomy, then, is derivative of their sense of responsibility and of their responsible conduct as rulers, not the other way around. Also, there are many examples of rulers who behave politically responsibly while being religiously quite heteronomous. Finally, autonomy does not call for responsibility except on a voluntary basis: once a person agrees to be a representative, he or she is responsible to their peers. But as far as autonomy is concerned, their services are not required. The society of fully autonomous enlightened individuals was deemed to be in no need of leaders anyway.

Leadership is needed. It is needed in order to lead, coordinate and arbitrate – as well as in order to show initiative in moments of stress. In the highly specialized world the national leadership may be divided between military leaders, economic leaders, industrial leaders, financial magnates, etc. This may give the impression that political leadership is no longer needed. Yet the coordination between expert leaders matters more, as it includes such decisions as to declare war or to send foreign aid. And it must remain in the hand of the national political leadership. This may be the leadership of the super-expert, the philosopher king, or it has to be flexible and democratic.

Who should lead? Popper objects to the question, as it invites the answer, the best, the philosopher king. The objection can be overruled, as the answer is both true and quite innocuous when coupled with Popper's own skepticism: we do not know who is the best. How then, do we decide? It does not matter, says Popper, if we can get rid of a bad ruler. This hides the supposition that rulers have to do little in order to qualify. This renders Popper's view conservative – a variant of classical liberalism. It is no longer correct in a world beset with todays problems and dangers.

Popper is aware of the new situation and the new problems it gives rise to. He has a proposal as to how to meet it. He offers it in the form of a maxim: *sagesse oblige*, in parallel to *noblesse oblige*. *Noblesse oblige* was at the end the feeble excuse the nobles had for their defense of their privileges: nobility carries more obligations than rights, they said; they were relieved of both despite their protests. *Sagesse oblige* may likewise be the technocrat's excuse. Now knowledge carries obligation, perhaps, yet by Popper's own lights we should not allow the knower to rule by virtue of his knowledge rather than by virtue of our democratic consent and under our democratic control. Popper's demand from the scientists and technologists is not that they rule, but that they study the unintended consequences of the implementation of their own ideas – especially the undesirable consequences. It seems clearly not wise to impose such a demand: the public might institute these studies to the extent that public responsibility requires them. If one objects to this alternative on the ground that the public needs leadership, then one thereby implies that with the maxim *Sagesse oblige* Popper has recommended that we appoint scientists as non-parliamentary political leadership. Admittedly, we need political leadership, both parliamentary and non-parliamentary; but we should not appoint the scientists to positions of leadership before asking the very question which Popper rejects: who should lead?

There is no reason to assume that Popper has recommended the appointment of scientists to any position of leadership. Rather, what he recommends is a sort of political self-appointment or, to use sociological terminology, political self-selection. This, however, is questionable. To some extent this self-selection is a matter of professional ethics – of technoethics (Mario Bunge) – which should evolve and should become obligatory for all technological research and perhaps even for all scientific research with immediate technological relevance. Yet political responsibility goes beyond that and so should not be obligatory, nor should it remain within any social group.

The politically responsible is selected, and he may be self-selected. In democracy and among spiritual leaders of many cultures this is so. The self-selected for political responsibility is a leader even when he does not hold an office. The one who does hold an office, on the contrary, is not always a leader. Yet, it is not enough to be willing – one has to be able. This is in part a matter of training, a matter of acquiring certain techniques. The training is usually acquired by apprentices, as in most crafts. In a technologically advanced society, the apprenticeship has to be accompanied by studies. These are absent these days since leadership still sounds undemocratic, and since to a large extent leadership cannot be learned since it is a matter of developing initiative. What is to be done about this?

The techniques of democratic leadership, such as moderating debates well, such as participating in committee activities, such as learning to make decisions and stand by them even while admitting in retrospect that they were mistaken – these must become popular knowledge. This is but the recognition of the pervasiveness of power politics in all spheres of life and in all organized society and sub-society. This is but the recognition that public participation in the democratic process is the best known way to secure and safeguard democracy (which can never be fully safeguarded). In addition to this, the very development of an intellectual framework for political studies which will transcend both the Enlightenment and the Romantic frameworks will open up much more room for the study of responsible leadership, and this knowledge can be disseminated and applied. The more we know what we require of a leader, the easier it will be to reconcile political leadership and democracy, and also to harmonize leadership with moral autonomy. It is time to have a society in which the lust for power is no longer a serious factor. It is, after all, not much different from the lust for duelling, once so very strong and now, happily, much diminished. It is, thus, not impossible (pace Adler and Russell) to reduce it to a manageable size.

5. CONCLUDING REMARKS

The problems of the modern world are more formidable than we ever faced. In this respect all opponents to technology are right. Yet a solution can only be found, if at all, by stepping technology up, not down, and by making it more rational, not less, as can be better achieved by democratization than by suppressing technocracy. Anyway, we should not relinquish responsibility to experts.

Perhaps it is too late. Perhaps the period of grace is practically over. It is nevertheless not impossible that rational swift action will change matters and make the difference, as Bertrand Russell said, between the destruction of humanity and the realization of some of its wild utopian dreams. These seem to be the only two options. The new free society of the wild utopian dreams will have to be more, not less, technological; more, not less, decentralized; more, not less, globally coordinated. It will have to live in an ecologically balanced world, as educated, limited in population, and peaceful. We can hardly manage with less than that. The struggle just may make all the difference between extinction and survival.

ENDNOTES

[1] It is not that possible emergencies are ignored. There is, for example, a document (available from the United States GPO Sales Program) prepared by the United States Nuclear Regulatory Commission, and its National Technical Information Service and published by the U.S. Nuclear Regulatory Commission and the Federal Emergency Management Agency in 1980, by the title of *Criteria for Preparation and Evaluation of Radiological Emergency Response Plans and Preparedness in Support of Nuclear Power Plants*, i.e., what should people in charge of nuclear power plants do in order to be prepared for a disaster in their plant. The most significant item concerns public education and information (pp. 49–51), and it says almost nothing: it lists a few current means of publications, such as posting information once in a while. Most of the document takes it for granted that in case of a disaster there will be no disruption of the social and the legal structure within which it might occur. This sounds reasonable, except for the case, included in the document (p. 1–10), of the possibility of a nuclear plant hit by a missile, or the one not mentioned – an earthquake. These three points will not go together: public information as described (and as happened in the Three Mile Island incident), a missile hitting a nuclear plant and social structures staying intact: many of the structures explicitly discussed in the document will have vanished from the vicinity in no time. Back to the drawing board!

[2] In all societies doubt is somehow allowed, since it happens anyway. But only in scientific societies the taboos on expressing it, and on asking impertinent questions, are (partly) lifted.

[3] See J. Agassi and I. C. Jarvie, editors, *Rationality*, Dordrecht, Martinus Nijhoff, 1986.

[4] Robert Boyle, the leading Fellow of the Royal Society, tried hard to conceal the fact that he was a member of the Sancta Sacraque Societas Cabalistical Philosophorum, a society devoted to alchemy, cabbalah and astrology. Astrology was ousted from the world of learning by Kepler, but the cabbalah and alchemy engaged Boyle and Newton. Alchemy never died. As the hope to transmute metals, alchemy was shared by many thinkers and it was officially made respectable by William Prout in 1815, and now enjoys the status of a scientific fact. See R. E. W. Maddision's life of Robert Boyle, London, 1969, p. 170.

[5] Cf. J. Agassi, *Towards a Rational Philosophical Anthropology*, The Hague: Martinus Nijhoff, 1977.

[6] Since external effects are not subject to the market mechanism, they must be controlled by other means, and the neo-classical economists have none except the normal democratic control by elections, which they know is not enough. A. O. Hirschman criticizes the neo-classical economic view as that which permits disgruntled consumers exist, i.e. disengagement, but not voice, i.e. no consumers' intervention in the market by means other than the market mechanism. No doubt, the more we center on consumer organizations, on workplace democracy and/or other innovative techniques of pluralist participatory democracy, the less relevant neo-classical economic theory is – even to the conduct of the market. Viewed this way, the impact of neo-classical economic theory, even in minor Keynesian modifications which it might suffer, is just this: neo-classical economics is a means for impeding the growth

of participatory democracy in the economy, and thus a defense of the status quo of the economic power-structure.

[7] Few statements are more popular in philosophy than the formula, Man is a Rational Animal. The formula is part of the Aristotelian theory of classification, which thinkers of the Enlightenment Movement regularly and increasingly distrusted. Yet the formula gained new significance.

[8] L. P. Williams, "The Politics of Science in the French Revolution", Paper Ten in Marshall Clagett, Ed., *Critical Problems in the History of Science*, Univ. of Wisconsin Press, Madison, 1959.

[9] Cf. Judith Buber Agassi, *Comparing the Work Attitudes of Women and Men*, Lexington Books, Lexington, Massachusetts, 1982.

[10] The fact that neo-classical economic theory is most insistent in its objection to all international trade barriers sounds inhuman in its indifference to the unemployment which the abolition of trade barriers may cause. The neo-classical economists have forceful responses to this change, some of which are interesting and even implementable with some reasonable modifications. But what the recommendation really amounts to is licence to big business and weapons industry to join ever so many local dictators in pilfering the tottering economies of their backward countries and supplying them with luxuries and sophisticated weapon systems rather than with the massive education including technological education and education for democracy which they need so badly.

[11] Marshall Macluhan's expression "global village" is curiously contrastable with Buckminster Fuller's "spaceship earth". The latter is concerned with the responsibility we should have towards the survival of humanity as a whole; the former with the fear of the utter uniformity and provinciality which may result from the highly desirable flow of all items cultural to all corners of the earth. The risk to human physical existence is all too real and can only be overcome by internationally coordinated proper legislation; the risk of global uniformity and provinciality cannot possibly be met by legislation; rather it can be met by a movement towards a pluralistic participatory democracy, where participation may be legally encouraged but not imposed.

[12] See my *Faraday as a Natural Philosopher*, 1971, p. 162.

[13] Experiments were performed – they are called Asch-type experiments – in which people are forced to choose between honesty and peer approval. Too often honesty loses. Still more painful is the fact that subjects may turn to pleading with their dishonest peers.

[14] There was an exception worth noticing: in the eighteenth century an economic theory existed which discussed the positive role of government for the economy as imposing protective import duties. But since Adam Smith has "proved" the inadvisability of such duties the debate died. To date classical economics is more individualistic political theory in economic garb than economics proper.

NAME INDEX

SUBJECT INDEX